Organic Food and Farming

Recent Titles in the

CONTEMPORARY WORLD ISSUES

Series

Media, Journalism, and "Fake News": A Reference Handbook
Amy M. Damico

Birth Control: A Reference Handbook
David E. Newton

Bullying: A Reference Handbook
Jessie Klein

Domestic Violence and Abuse: A Reference Handbook
Laura L. Finley

Torture and Enhanced Interrogation: A Reference Handbook
Christina Ann-Marie DiEdoardo

Racism in America: A Reference Handbook
Steven L. Foy

Waste Management: A Reference Handbook
David E. Newton

Sexual Harassment: A Reference Handbook
Merril D. Smith

The Climate Change Debate: A Reference Handbook
David E. Newton

Voting Rights in America: A Reference Handbook
Richard A. Glenn and Kyle L. Kreider

Modern Slavery: A Reference Handbook
Christina G. Villegas

Race and Sports: A Reference Handbook
Rachel Laws Myers

World Oceans: A Reference Handbook
David E. Newton

First Amendment Freedoms: A Reference Handbook
Michael C. LeMay

Medicare and Medicaid: A Reference Handbook
Greg M. Shaw

Books in the **Contemporary World Issues** series address vital issues in today's society such as genetic engineering, pollution, and biodiversity. Written by professional writers, scholars, and nonacademic experts, these books are authoritative, clearly written, up-to-date, and objective. They provide a good starting point for research by high school and college students, scholars, and general readers as well as by legislators, businesspeople, activists, and others.

Each book, carefully organized and easy to use, contains an overview of the subject, a detailed chronology, biographical sketches, facts and data and/or documents and other primary source material, a forum of authoritative perspective essays, annotated lists of print and nonprint resources, and an index.

Readers of books in the Contemporary World Issues series will find the information they need in order to have a better understanding of the social, political, environmental, and economic issues facing the world today.

CONTEMPORARY WORLD ISSUES

Organic Food and Farming

A REFERENCE HANDBOOK

Shauna M. McIntyre

An Imprint of ABC-CLIO, LLC
Santa Barbara, California • Denver, Colorado

Library of Congress Cataloging-in-Publication Data

Names: McIntyre, Shauna M., author.
Title: Organic food and farming : a reference handbook / Shauna M. McIntyre.
Description: Santa Barbara, California : ABC-CLIO, [2021] | Series: Contemporary world issues | Includes bibliographical references and index.
Identifiers: LCCN 2020033440 (print) | LCCN 2020033441 (ebook) | ISBN 9781440870033 (hardcover) | ISBN 9781440870040 (ebook)
Subjects: LCSH: Organic farming. | Natural foods.
Classification: LCC S605.5 .M347 2021 (print) | LCC S605.5 (ebook) | DDC 631.5/84—dc23
LC record available at https://lccn.loc.gov/2020033440
LC ebook record available at https://lccn.loc.gov/2020033441

ISBN: 978-1-4408-7003-3 (print)
978-1-4408-7004-0 (ebook)

25 24 23 22 21 1 2 3 4 5

This book is also available as an eBook.

ABC-CLIO
An Imprint of ABC-CLIO, LLC

ABC-CLIO, LLC
147 Castilian Drive
Santa Barbara, California 93117
www.abc-clio.com

This book is printed on acid-free paper ∞

Manufactured in the United States of America

Organic farming has grown from a fringe community to a booming business with billions of dollars at stake. Nearly every food retailer in the United States carries at least some organic food products, and most Americans buy organic food occasionally. Why has it grown to be such a popular commodity? Many people believe that organic food is better for themselves or the environment, but very few can actually describe exactly what makes it organic. Plenty of people question if it is really better or if it is just a marketing ploy. These questions and many more are answered in this book.

In an era of fake news and misinformation, it can be hard to determine what is fact and what is fiction. Everyone has to eat. I believe that everyone should be able to understand what choices they are making at the grocery store and how those choices impact others and the earth. I wrote this book because I believe that having access to good information is necessary to make good choices. This book draws on academic and peer-reviewed sources to provide a balanced overview of organic food and farming and a starting point for finding other good sources of information. I cover ways in which organic fails to meet expectations alongside the ways in which it is successful. I cover the issues that most commonly land in the headlines and those that impact the farming community the most. Organic farming came about because a few individuals were dissatisfied with the scientific and commercial direction that farming had taken. Others joined as they tried and failed at making chemical farming a success. This book provides a comprehensive overview of the growth of organic both as an industry and a social movement and the inherent challenges that occur from trying to be both. It covers controversies such as hydroponics, the dairy crisis, GMOs, and labeling fraud. It discusses the role of farmers, processors, and retailers in the growth of the industry and how it has evolved since the development of a federal certification program. It goes beyond that to discuss how farmers continue to struggle with the tension between certification and the philosophical underpinnings of organic agriculture.

Chapter 1 begins with a short overview of the history of agriculture and describes the origins of organic farming. It delves into the historical context that led to the development of an "alternative" agriculture including the social, economic, technological, and political issues that drove it forward or held it back. It follows the creation of federal regulation and support for organic farming and the impact it had on the organic farming community and food production sector. In a matter of a couple decades, organic food went from being a tiny, almost invisible, segment of food sales to a booming industry with global reach. The first chapter looks at how organic farming has become both mainstream and a part of the local alternative food networks that have been established across the country.

Chapter 2 begins by presenting the current evidence on the most common questions and myths about organic farming and food, including the impact on human health and the environment. I explore the role that organic farming might play in addressing the climate crisis and whether it can extend its reach into social justice issues. Next, I describe several crises and controversies occurring in the organic sector and what steps might be taken to address these problems.

Chapter 3 offers up a wonderful collection of essays from farmers, academics, and others in the organic sector who share their own insights and perspectives on topics such as research, GMOs, and farming challenges.

Chapter 4 profiles a number of fascinating individuals who helped form and shape the organic movement from its beginnings to the present day. It also includes a section on organizations that have been advocating for and supporting the organic industry from its earliest beginnings in the United States.

Chapter 5 is a data and documents chapter that presents recent industry statistics and key government and legal documents. Chapter 6 is a comprehensive resources list with short summaries describing each source. This is a good place to look for more detailed information about a particular topic.

Chapter 7 contains a detailed chronology of the key events in the history of the organic sector from 1840 to the present. Finally, the glossary at the end of the book defines a number of words commonly used when discussing issues in organic food and farming.

Organic Food and Farming

Introduction

Organic food can be found in grocery stores all over the world. It makes headline news on a regular basis. But what is organic food and how did it grow to be a huge industry? This chapter will discuss the history of agriculture and what gave rise to the development of organic farming. Technically, agriculture is the art and science of cultivating the land, harvesting crops, and managing livestock, while farming is the business of running a farm. In this book, agriculture and farming are used interchangeably to reflect how they are commonly used. The chapter will explore the social, economic, and political circumstances that laid the groundwork for modern organic agriculture and will discuss a number of definitions of organic and the farming practices that are used in organic farming.

The chapter will follow the development of the organic movement from its earliest beginnings in the 1940s to the present day. In doing so, it will trace the creation of a social movement, a booming industry, and a federal regulatory program. In the United States, organic farming began as a farmer-driven interest that spread in popularity along with the rising concern in environmental issues. As interest in organic food began to spread, the need for policy intervention became apparent. This chapter details the steps that were taken in creating state and

---

An organic farmer shows off her freshly harvested carrots. Organic farming is often more labor intensive than conventional farming. (Michael Walters/Dreamstime.com)

federal legislation to regulate and eventually support organic food production. Finally, the chapter concludes with a summary of how the organic industry has changed since the creation of a national organic program.

A Short History of Agriculture

In order to understand organic agriculture as it exists today, you must understand the social and economic forces that created it. Organic agriculture is both an age-old practice and a modern invention. Many argue that for much of human history, all agriculture was organic. It wasn't until synthetic chemicals came into widespread use in the 1920s to 1940s that any other method of agriculture existed. There has always been a wide variety of approaches to agriculture, both philosophically and practically. A brief look at the history of agriculture will provide a clear context for the evolution of farming systems and why organic agriculture came to be a booming and federally regulated industry, not just a farming method.

Early Development

Agriculture's earliest forms began somewhere in the range of 8,000–11,000 years ago (Leigh 2004, Standage 2009). For much of the history of agriculture, farming was done in small plots using human labor with the goal of feeding a family or a village. Many of these were closed systems that used animal manures to build soil fertility, but not all farming was done sustainably. Often, early farming systems used flood plains, or slash and burn techniques, or incorporated a fallow rotation that rejuvenated the soil without much human effort or intent (Barton 2018, Leigh 2004). As societies became more sophisticated, some early agricultural technologies were developed including irrigation systems, manure fertilizers, salt or lime applications, crop rotation, plowing, terracing, and seed selection (Leigh 2004). Not much changed until an era of rapid scientific discovery occurred from the late middle ages to the late 1800s. Many of these discoveries had a great impact on the philosophical and technological approaches to agriculture. This was also a time of great population growth and exploration. Many Europeans emigrated to the Americas and saw the vast expanse of land as a resource to be used without regard for the future. This concept, paired with the enthusiasm with which society embraced scientific advances, led to our modern agriculture systems.

The Introduction of Scientific Concepts

In the mid-1800s, soil science and organic chemistry research had taken off with the publication of Sir Humphrey Davy's book *Elements of Agricultural Chemistry* (Barton 2018, Leigh 2004). At that time many believed in the humus theory, wherein plants derived their nutrients from water-soluble compounds of carbon, hydrogen, oxygen, and nitrogen contained in humus (Heckman 2006). It was also believed that plants could generate other nutrients from these four essential elements. Although many people were adding salts or lime to the soil by this time, they believed it was to help break down the humus rather than to add new nutrients. Carl Sprengel, a German chemist, conducted numerous analyses of plant matter and humus extracts and found out that a variety of salts such as nitrates, sulfates, chlorides, magnesium, and phosphates among others were in fact feeding the plants. He essentially disproved the humus theory and laid the groundwork for the law of the minimum and the creation of fertilizers (Heckman 2006, van der Ploeg et al. 1999).

In 1840, Justus von Liebig built on the work of Carl Sprengel to articulate the law of the minimum, which states that plant growth is determined not by the total resources available, but by the scarcest resource. Liebig used this law to hypothesize that inorganic materials could provide plants with the required nutrients just as effectively as organic sources, such as animal manures. He went on to break down the essential nutrients needed by a plant and argued that not all nutrients were needed equally for plant growth. This led to his development of a nitrogen-based fertilizer. He published the book *Organic Chemistry in Its Application to Agriculture and Physiology*, which had profound impacts on the production of agriculture and the use of synthetic fertilizers that continues today (Barton 2018; Heckman 2006; Leigh 2004; van der Ploeg et al. 1999). Soil fertility was not the only problem that farmers sought to solve by adding substances to plants. Sulfur, copper, and arsenic were among the first pesticides used to prevent damage from fungus and insects (Barton 2018).

By the late 1800s, the Industrial Revolution was underway, and agriculture evolved along with it. Farmers and scientists began experimenting with genetics and plant hybridization. As new technology, such as tractors, grain silos, and threshing machines, became available for widespread use, large-scale monocultures became the norm. By the early 1900s, many farmers in North America and Europe were rapidly shifting to a chemically and technologically intensive approach to farming. This increased the instances of depleted soil fertility and crop failures that led to famines (Standage 2009). In the 1930s Dust Bowl era, the U.S. Midwest experienced severe loss of topsoil and large dust storms due to the mechanized methods of farming and lack of attention to soil fertility.

In the 1920s, access to affordable chemical fertilizers and pesticides created a new approach to agriculture. Farmers began experimenting with by-products of coal or other industrial by-products. The Haber-Bosch process, developed in the early 20th century, was a method of converting hydrogen and nitrogen to ammonia, creating a bio-available form of nitrogen on an industrial scale (Leigh 2004; Paull 2009). The industrial production of ammonia was originally started in Germany and used in World War I to create ammunitions. Once the war was over, the cheap production of synthetic nitrogen-fixing ammonium spread across Europe and North America and was marketed to farmers as a way to increase yields.

By the 1940s, Dichlorobiphenyl trichloroethane (DDT) and organophosphates were found to be effective pesticides

(Davis 2014). Synthetic herbicides were also developed during this time with 2,4-dichlorophenoxyacetic acid (2,4-D) being the most popular (National Research Council 2000). Much of this research took place in England at Rothamsted Research station located in England, one of the oldest agricultural experiment stations still in existence. After World War II, chemical farming became the norm until persistence and toxicity of the chemicals began to take heavy tolls on environmental and human health.

Then in 1962, Rachael Carson sounded the alarm with her book *Silent Spring*. Eventually, concerns about pesticides led to the creation of the Environmental Protection Agency (EPA) and a ban on DDT and other organochlorine compounds. Synthetic pesticides, including various forms of herbicides, fungicides, and insecticides, have continued to be used widely, although they must go through an approval process with the EPA before being sold to farmers (Davis 2014). In the 1990s, genetic engineering introduced new methods of pesticide control in combination with synthetic organic compounds. Crop seeds were genetically modified to be resistant to herbicides so farmers could apply glyphosate-based herbicides without damaging the crop plant (Davis 2014).

At the same time that synthetic chemicals were changing crop production, farm structure and animal production were undergoing their own transformation. Shortly after World War II, scientists discovered that incorporating broad spectrum antibiotics in grain feed given to poultry, cattle, and pigs boosted their growth rate while using less grain feed. Animals reached their market weight much faster and the side benefit of reduction in diseases meant farmers could raise animals in large numbers confined in barns or feedlots (Gustafson and Bowen 1997; Kirchhelle 2018). Growth hormones such as rBGH were incorporated into livestock production starting in the 1950s and are now commonly used by dairy farmers to increase milk production (Dibner and Richards 2005). The expanded production that occurred on increasingly specialized farms led to the consolidation of food processing and distribution and eventually vertical integration or consolidated ownership of many aspects of agriculture and food production (Ikerd 2016).

The Rise of Large-Scale Farming

Between 1935 and 1974, the number of farms in the United States decreased from 6.8 million to 2.3 million, and it has since leveled off (U.S. Department of Commerce 1978). At the same time, the size of individual farms has been steadily increasing from an average of 100 acres in the 1930s to more than 400 acres today (USDA ERS 2017). Commodity prices have remained relatively the same because of overproduction, but the cost of implementing new technology has increased dramatically (Conkin 2008). This has led to a host of challenges for the farming sector, including huge debt loads, rural to urban migration and the emptying of small rural towns, and loss of farmer-controlled agriculture production (Ikerd 2016). The rapid changes that occurred in agriculture over the past century have generally occurred as science, policy, and economic structures have favored technological solutions and approaches that emphasize increased yields and production over everything else. In a relatively short span of time, agriculture went from diversified farms that were largely dependent on local farm-level cycling of nutrients to a reliance on off-farm inputs. The heavy use of inputs created a more specialized and standardized form of agriculture, where livestock and crop production were separated and isolated (Foster and Magdoff 1998). As a whole, agriculture was seen less as a management of natural cycles and more as an industrial model of production. Organic agriculture's adoption and growth has largely been a response to the changes that have occurred since the Industrial Revolution. Those who favored organic agriculture and other alternative methods of production often looked for more ideological solutions to the challenges in farming, ones that focused on the ecology of the soil and integration of natural nutrient cycles. Others wanted to avoid the large debts inherent in industrial agriculture and to maintain more independence in their farming life. As organic agriculture has matured, many of the same issues that have plagued industrial agriculture have begun to concern organic farmers, and the result is the development of even more alternatives including the recent interest in regenerative agriculture.

What Is "Organic Agriculture"?

Organic agriculture as a distinct method of food production began in the early 1900s as a reaction to the industrialization of agriculture and the pressing problems that resulted. What sets organic farming apart from other sustainable or alternative systems of production is the fact that it has become clearly defined and regulated. Organic agriculture is the only farming system that is regulated by laws in many countries and signified with a legal certification (Seufert et al. 2017). As of 2017, 87 countries have some form of organic standards or are in the process of developing them, and nearly every country has some organic production (IFOAM 2018). Many countries have created bilateral agreements to facilitate trade and ensure regulations meet import rules. While it is true that to some extent organic agriculture has a clear definition that is determined by legal codification, the reality is that there is a broad spectrum of organic production systems that are only marginally related (Rigby and Cáceres 2001).

Defining Organic

Modern organic agriculture is a blending of many different techniques and is hard to quantify, though there are many attempts to do so. Some of the terms commonly used to describe organic production include: integrated or holistic systems; environmentally and economically sustainable management practices; reliance on on-farm resources; understanding and management of ecological and biological cycles; and an emphasis on healthy soils, plants, and animals.

For some, organic farming is a set of values, principles, and ideologies along with specific farming strategies to enhance the soil and protect the environment. For others, organic is a set of rules and guidelines specified by regulations under the National Organic Program (NOP) administered by the United States Department of Agriculture (USDA) and similar regulating bodies around the world. There is much overlap between the two versions of "organic," but the end results are often very different. The idea that the way something is produced is as important as the final product underpins the philosophical definition of organic (Clark 2015). The philosophical definition of organic incorporates social, economic, and environmental principles. The difficulty comes when trying to translate those principles into specific production practices. There is a wide range of interpretations on how to farm and which technical aspects of production best include the values of organic agriculture. The interplay of principles and practices is complicated at best and often contentious at worst. Farmers who want to sell their products as organic need guidelines on how to produce organic food and consumers need some assurances that their understanding of organic food meets their expectations. The more distance between a food producer and consumer, with processing, distributing, and retailing in between, the more complex nodes of separation exist, through which the translation of values can get lost. The process of codifying organic farming then attempts to translate a set of values into specific criteria of allowable techniques. The larger the scale, the more difficult it becomes to maintain the social and environmental principles inherent in the organic agriculture movement. These values are not enforceable or traceable unless reduced to a set of tangible criteria. Since most consumers are interested in the health aspects of organic food including the lack of pesticides or synthetic additives and genetically modified organisms (GMO), those are typically the factors that get the most attention in determining what is certified organic.

Before the term "organic" was coined, Sir Albert Howard, a pioneer of the farming system, described it like this, "Mother earth never attempts to farm without livestock; she always raises mixed crops; great pains are taken to preserve the soil and prevent erosion; the mixed vegetable and animal wastes are converted into humus; there is no waste; the processes of growth and the processes of decay balance one another; ample provision is made to maintain large reserves of fertility; the greatest care is taken to store the rainfall; both plants and animals are left to protect themselves from disease" (Howard 1940, p. 4). So, at its origin, there was an emphasis on mimicking natural cycles, viewing the farm as an integrated or whole system with little to no outside inputs. Organic agriculture is not a return to old ways of farming as many of the historical systems of farming were not organic at all (Barker 2016), but it does relate to a time in which people were more connected to the cycles of the earth, a mindset that was eroded during the Industrial Revolution. In fact, many people suggest that organic farmers are inherently opposed to progress and scientific research. The truth is that organic farmers are continually testing new techniques and looking for better methods of production. There are some who try to relate the definition of organic agriculture to the use of the term "organic" in biology or chemistry. In biology, the term "organic" means derived from living organisms, and in chemistry, it indicates carbon-containing compounds (with some exceptions). Using those limited definitions would make one miss many of the practices that derive from a whole system view and limit useful production practices (Barker 2016). Organic farms are highly managed systems with lots of integrated parts. Successful organic farming requires lots of observation, adaptation, and record-keeping. The most basic element of organic philosophy and practices comes down to the mantra: feed the soil, not the plant.

A major commonality across regulations from around the world is that organic is defined as an agricultural management system with heavy emphasis on avoiding synthetic inputs over reliance on natural processes. The codification of organic agriculture has distanced organic production models from the foundational goals of the organic movement such as improved water quality, soil fertility, and biodiversity (Seufert et al. 2017). Early organic farmers sold their products to local consumers and built up their reputation with personal connections. As organic food grew in popularity and organic production grew to meet the demand, farmers needed some way to market and prove that their products were grown organically. This led to the development of certifying agencies and organizations and eventually state and federal regulations that defined what could be sold as organic food. The move to codify organic agriculture was largely driven by consumers, which means the perceived health benefits of organic in the form of fewer chemicals on food are emphasized over the holistic management practices that get missed in the reductionist take on regulations (Seufert et al. 2017).

As agriculture in general started to incorporate more technical solutions, defining organic agriculture often became about what it was not, rather than what it was. When the USDA's *Report and Recommendations on Organic Farming* was written in 1980, the study team created a definition from their findings that reflected the range of farming practices they discovered through case study interviews. It stated, "Organic farming is a production system which avoids or largely excludes the use of synthetically compounded fertilizers, pesticides, growth regulators, and livestock feed additives. To the maximum extent feasible, organic farming systems rely upon crop rotations, crop residues, animal manures, legumes, green manures, off-farm organic wastes, mechanical cultivation, mineral-bearing rocks, and aspects of biological pest control to maintain soil productivity and tilth, to supply plant nutrients, and to control insects, weeds, and other pests" (USDA Study Team 1980, p. 9). This definition was clearly focused on the specific practices that were and were not used by organic farmers rather than the principles behind those practices.

Thus, certified organic was defined in terms of opposition to those things that organic farmers and advocates agreed should not be a part of agriculture. Organic farming was defined by the fact that it does not make use of synthetic fertilizers; GMOs; treated seeds; irradiation; and chemical pesticides, herbicides, or fungicides. According to the USDA NOP, organic is "the application of a set of cultural, biological, and mechanical practices that support the cycling of on-farm resources, promote ecological balance, and conserve biodiversity. These include maintaining or enhancing soil and water quality; conserving wetlands, woodlands, and wildlife; and avoiding use of synthetic fertilizers, sewage sludge, irradiation, and genetic engineering. Organic producers use natural processes and materials when developing farming systems—these contribute to soil, crop, and livestock nutrition, pest and weed management, attainment of production goals, and conservation of biological diversity." Thus, organic certification is a labeling term that signifies a product has been grown and handled in accordance with the USDA standards and certified by an accredited agent. To become certified, a farmer must be free of prohibited substances for three years, create an organic management plan for the farm, and submit to annual inspections. This is part of an application process that includes a certification fee.

Though the legal definition is often distilled down to the aspects used as a legal definition of organic production for certification processes, those focused primarily on a prescribed set of production practices, the National Organic Standards Board (NOSB), did recommend a wider definition of organic. The advisory board made a formal recommendation to the USDA to include this definition of organic in their final rule. It states that "certified organic agriculture is an ecological production and management system that promotes and enhances biodiversity, biological cycles, and soil biological activity based on management practices that restore, maintain, and enhance ecological harmony and minimal use of off-farm inputs" (NOSB 2000).

A number of other terms are often used when describing organic farming including naturally grown, biologically grown, ecologically grown, chemical-free, sustainable agriculture, and so on. The term "organic" stuck because a few thought leaders continually used it to describe the set of principles and practices they were advocating. It was the term that caught on and eventually came to have a codified set of standards associated with it. Many of these other terms came into wider use when the term "organic" came to be viewed as polarizing and divisive, including the term "sustainable agriculture" that is still widely used today. Using terms other than organic often made the farming practices associated with organic agriculture more politically palliative to researchers, farmers, and policymakers.

Organic Practices

While academics are describing organic agriculture as a social movement, many farmers choose it, not for philosophical reasons, but for practical reasons, and businesses use it to tap into the growing market share. Although many organic farmers do feel strongly about the philosophical elements of organic farming, there are many others who think it just makes sense or is a good business decision. Many farmers cite economic reasons for their motivation to farm organically (Goldberger 2011). Others are concerned with the risks of farming with chemicals. Pesticide poisoning of farmworkers is a significant cause of death and one that is mitigated with organic production practices. While organic farming is often described in terms of how organic is different from conventional, such as not using chemicals, GMOs, or growth hormones, organic farming is much more than that to the farmers who follow its practices.

Organic farming requires a broad set of skills and a strong understanding of local ecological systems to develop proper management practices on the farm. If left alone, nature follows a cycle of growth, death, and decay with rodents, earthworms, insects, and microbial organisms in the soil doing the work of recycling nutrients. In agriculture, by necessity, the cycle is disrupted, and that disruption requires farmers to manage pests, nutrient cycles, and more. Conventional farmers draw on chemicals and genetically modified crops to manage their farms, while organic farmers choose to grow food in ways that encourage biological diversity and, to the extent possible, mimic natural cycles and balances. In practicality, this requires organic farmers to first observe the cycles present on their land and then develop systems to support those cycles. Organic farmers use a number of techniques to achieve good crop yields while addressing their biggest challenges. Organic farmers must tailor their practice to their farm type, size, and location, but what they all have in common is their reliance on diverse crop rotations, organic fertilizers, natural pest management, and use of organic feed for livestock (Duram 2005).

Composting is a critical element of organic farming and key to maintaining healthy balanced soils without resorting to chemical fertilizers (Coleman 1995; Martin and Gershuny 1992). While the practice of composting is in fact ancient, significant research on composting has advanced our understanding of how to create a good balance of nutrients and micro-organisms needed in the soil. Application of appropriate compost can improve soil texture, structure, moisture-holding capacity, and control erosion. It also improves aeration and makes nutrients more accessible to the plant (Martin and Gershuny 1992).

For many organic farmers having an integrated farm with both crops and livestock is vital to producing a cycle of nutrients. Farmers raising livestock maintain living conditions that support good health and natural behaviors. Most animals require time outside on pasture and minimal confinement. The livestock are not given growth hormones or antibiotics unless medically required. Similar to crop production, there are cultural, biological, and mechanical methods of management. An example of cultural management would be selecting a breed suited to climate and terrain. A biological example would be to rotate pastures for maintaining weeds and preventing parasites. A farmer might rotate sheep, cattle, and chickens through a section of pasture to reduce parasites or other pests, and grazing livestock can have symbiotic benefits on an organic farm. Livestock gain a varied nutrient-dense diet and fields have reduced weeds, improved fertility, gleaning of unused crops such as apples that may rot, and much more. For many farmers, incorporating livestock and crops (fruit, vegetable, grain, etc.) creates a diversified income stream that can improve income and insulate against challenges such as market downturns or drought. Organic farming is also very scale-dependent. While some types of organic operations do well to operate at a large-scale such as grain production or grain and livestock mix, others have significant challenges when scaled up, and the impacts on the environment often increase along with scale, especially when organic production mimics monocultures and feedlot-style livestock production.

Some of the biggest challenges for organic farmers are pests, weeds, and diseases. Organic pest management draws on a number of creative techniques developed over many years. Strategies to prevent pest problems by maintaining a healthy balanced ecosystem on the farm are critical for avoiding chemical pesticides. Selecting plants that are naturally pest resistant and ensuring they are healthy is considered a cultural control. Timed planting, crop rotation, and inter-cropping with green manure crops are all methods organic farmers use to simultaneously keep soils and plants healthy while discouraging an infestation of pests (Coleman 1995). A biological control would include beneficial insects, which do everything from pollinating plants to eating pest insects and breaking down decaying organic matter. Farmers can improve the number of beneficial insects and animals by creating good habitats. For example, establishing what are called companion plants in the borders of a field can induce a number of beneficial insects to the farm. Biological controls are another method farmers can use to control pests. These can be anything from dusts and sprays of fungus, bacteria, or parasites, which target a variety of pests and predatory insects (Ellis and Bradley 1996). To control weeds, farmers use a range of covers including plastic, fabric, and mulch (Gomiero et al. 2011).

Organic farmers rarely rely on just one technique to manage their crops and livestock. They are constantly refining their approach and changing it over time as growing conditions change due to changes in climate and as their knowledge of the land improves. Farmers have a long history of testing new techniques and sharing their knowledge with one another since modern research has largely ignored their needs. Farmers also shift their methods to match changing market conditions, personal needs, and availability of seeds or other inputs. As the organic industry has evolved, many of the techniques used to farm organically have evolved along with it.

The Origins of the Organic Movement

The history and growth of the organic sector in the United States can be characterized by three distinct eras of growth. Each era of the movement was driven by external social and economic factors. The first era of the organic movement began against the backdrop of growing scientific discoveries and chemical agriculture around the 1940s and 1950s. Organic pioneers were looking for ways to keep up with the desire to improve agricultural yields while maintaining most elements of their traditional agricultural practices. This was a time of experimentation and independence by organic farmers, and there was often little contact between them. The subsequent era was framed against the backdrop of the environmental movement and the back-to-the-land culture that came about in the 1960s and 1970s. As interest in organic agriculture grew, so did efforts to share knowledge and to connect communities of organic growers and consumers. This era also saw some significant pushback to the organic movement by researchers, government officials, and conventional agribusinesses.

Finally, beginning in the 1980s and continuing today, the organic industry found significant mainstream business growth and government support. This launched the rise of organic legitimacy in the form of research, policies, certification, and a fully formed organic sector, albeit one that is similar to the conventional sector in many ways. Today, the organic industry resembles very little of the pockets of organic farming communities of the early days. Organic food now comes in a full array of products and is found in most retail food outlets. Organic farmers no longer need to rely on their own trial and error, but they have many more resources to draw on for information about farming and marketing organic food. That said, there is still a long way to go before organic is supported and legitimized in the same way that conventional agriculture is supported by government agencies, researchers, and the general public.

The pioneers of organic agriculture were largely from Britain and Europe, but there were a few individuals in North America and elsewhere who had a significant impact on the understanding of organic production methods. F.H. King, an agricultural scientist and professor from Wisconsin, toured Asia and wrote about the agricultural practices he saw there, including the recycling of waste material to improve soil fertility, in his book, *Farmers of Forty Centuries, Permanent Agriculture in China, Korean, and Japan*, published in 1911. In the 1920s and 1930s, Austrian, Rudolf Steiner and German scientist, Dr. Ehrenfried Pfeiffer developed an approach to farming that viewed the farm as a whole living system, something they called biodynamic farming. The work was popularized through the book *Bio-dynamic Farming and Gardening* written by Dr. Pfeiffer in 1938. Their work influenced the thinking of a British agriculturalist, Lord Northbourne, and an American publisher, Jerome Irving (J.I.) Rodale. In 1939, Lord Northbourne hosted a summer school with Dr. Pfeffer and adopted the biodynamic principles of farming on his own property (Paull 2011).

In 1940, Lord Northbourne published the book *Look to the Land*, in which he coined the term "organic farming." Around the same time, Sir Albert Howard, an economic botanist and researcher, published *An Agricultural Testament*, a foundational text on composting, soil fertility, and the practicalities of growing organically, although he did not as yet use the term "organic." Albert Howard spent 30 years as a government scientist in India, where he developed the Indore system of composting, a foundational aspect of organic farming. Albert Howard first adopted the term "organic" when he published *Soil and Health: A Study of Organic Agriculture* in 1945. His work inspired and influenced many others who were starting to look for alternative systems of agriculture.

In Britain, Eve Balfour used the term "organic" in her book *The Living Soil* published in 1943. She brought together a group of people to form the Soil Association in 1946 and began one of the longest running comparison studies of organic and conventional agriculture. While Balfour was inspired by the work of Albert Howard, he did not support her efforts in creating the Soil Association. Even among the early pioneers, there was plenty of disagreement about what constitutes organic. Howard was opposed to anything that strayed from a strictly scientific approach, while Balfour embraced many of the more spiritual elements derived from the biodynamic approach.

In North America, J.I. Rodale, a publisher of health magazines, read books by Howard and King and used them to help frame his own ideology of farming. After corresponding with Albert Howard, Rodale began publishing the *Organic Farming and Gardening* magazine in 1942. In the magazine, he wrote about the work of Howard and others and served as the main promoter of organic practices in the United States for many years. He went on to create the Rodale Institute to support organic research. He published *Pay Dirt* in 1948, one of the first books to link chemical farming with health. The ideas were not necessarily new concepts, but they were popularized and given some measure of legitimacy by these new publications coming out in the 1940s. Although there was a small following of farmers interested in the work of these pioneers, there was no concerted effort to teach or spread these ideas within the United States.

It was J.I Rodale's interest in Albert Howard's work that really sparked a larger movement. Rodale invited Howard to be an associate editor for his magazine, and the two of them wrote numerous articles about the Indore method of composting and other organic methods. Interest in the magazine grew, and others began trying out the organic farming methods. While not specifically part of the organic movement, two individuals, Henry Wallace, first U.S. Secretary of Agriculture and then Vice President, and William Albrecht, a professor of soil science at the University of Missouri, were encouraging farming techniques such as integrating forestry and farming, returning organic matter to the soil, and linking soil fertility with that of human health. This provided the beginnings of academic work on organic agriculture, although it took many years to be called that. Much of Albrecht's work was published in ACRES USA, a publication founded by Charles Walters, Jr., who made it his mission to encourage organic production for large-scale farmers and the use of alternative soil amendments produced by a growing input industry (Youngberg and DeMuth 2013).

Growth of the organic movement was slow through the 1950s, but like many other issues related to the environmental movement, Rachel Carson's *Silent Spring* published in 1962 caused a huge awakening to the effects of pesticides and herbicides used in agriculture. The resulting health concerns and pushback to corporate-driven agriculture corresponded with other significant social movements including the counterculture back-to-the-land movement. Increasing numbers of nonrural people were suddenly interested in organic farming. By the end of the 1960s, Rodale's *Organic Farming and Gardening* magazine had over a million subscribers (Barton 2018).

The environmental movement motivated Congress to pass numerous environmental laws including the Pesticide Control Act (1972) and the Toxic Substance Control Act (1976) and to develop the EPA. As the interest in organic farming and food grew, so did the opposition from the conventional food industry. Pushback to organic agriculture during this era was intense and openly hostile. The Secretary of Agriculture, Earl Butz, made several infamous pronouncements against organics and set the tone for interactions between organic farmers and government staff.

The main arguments against organic agriculture tended to fall into one of three categories: there is nothing special about organic because everything is derived from nature, even pesticides; the growing world population needs food and organic agriculture cannot produce enough; and finally, organic agriculture is based on myth and romanticism. These criticisms reached their peak in what was considered the age of science when the push to find technological solutions for everything was a driving social and economic force. Each of these criticisms had roots in the truth, which often made it hard to defend and created conflict within the organic community. Albert Howard, for example, distanced himself from those who embraced biodynamic practices because they contained elements of mysticism. Opposition to organic in the 1970s often related to agricultural scientists' and researchers' memories of hard work on the mixed farms of their youths coupled with a desire to embrace technological solutions and the latest great scientific discoveries (Youngberg and DeMuth 2013). Other public figures were supporting organic production methods, including Wendell Berry, Jim Hightower, and Barry Commoner, who used their writing to elevate the organic movement (Obach 2015).

Farmer Organizations

As the number of organic farmers started to increase, they naturally wanted ways to connect and learn from one another. J.I. Rodale encouraged the creation of organic garden clubs and published a directory of clubs that formed around the country. By 1970, there were over 100 clubs located across North

America and published in the directory. In the early 1970s, organic farmers started to organize themselves into more formal local and regional groups. Farmers in Maine and Vermont were among the first to create formal organizations (Maine Organic Farmers and Gardeners Association and Northeast Organic Farmers Association) with a focus on farmer support in 1971, closely followed by Oregon Tilth in 1974.

Other organizations formed throughout the 1970s, including one in Michigan in 1973 and one in Ohio in 1979. Organic farming tends to be concentrated on the coasts and eastern portion of the Midwest, and that is where organizations typically began. It wasn't until the 1990s with the formation of the Upper Midwest Organic Farming Conference in 1990 and the subsequent creation of the Midwest Organic Sustainable Education Services (MOSES) in 1999 that there was any substantial organic training in the middle of the country. Many other organizations have since formed, and most states have at least one organization that offers support or training relevant to organic farmers.

In 1972, delegates from five organic organizations around the world met in France to discuss opportunities for advancing the organic movement and the need for scientific research to support the movement. The delegates, including one from the Rodale Institute in the United States, created the International Federation of Agricultural Movements (IFOAM). Despite the international gathering, a more cohesive organic movement was a long way off. The United States still had no national organizations representing the organic movement, and interaction between regional groups was minimal. The regional organizations were formed as a way to share knowledge and support new entrants to organic farming, and not to promote a wholesale shift to organic farming and create the support and infrastructure that would be necessary to achieve it.

There were a few regional grower organizations that had begun to address the demand for more growth and clarity around what should be considered organic by offering certification. In 1971, Rodale Press used their connections to the organic community to institute a set of criteria for organic practices and implemented the first organic certification system in North America. Then the California Certified Organic Farmers (CCOF) launched a certification program in 1973 with 54 farmers, and by 1975, there were 10 more organizations offering some form of organic certification. These early certification standards emphasized the philosophy of organic production over technical details, which led to many arguments over which practices were considered organic and which were not. Then in 1978, IFOAM began work on a set of baseline standards based on their "principles of organic," which included principles of health, ecology, fairness, and care. They were meant to set a standard against which other organizations around the world could base their own certification standards. It was many years before the United States had a unified set of standards for organic farming.

Creating a National Organic Program

The path to a federally mandated program for organic agriculture began in 1980 with the publication of a report on the status of organic farming. Secretary of Agriculture Robert Bergland was interested in how organic farming practices might be used to ease some of the challenges farmers faced from a farm crisis brought on by a cost-price squeeze from increasing debt, more competition from imported food, and higher energy costs. He put together a study team who conducted interviews, tours, and surveys, and reviewed and compiled the information into a document called *Report and Recommendations on Organic Farming*. Those in support of organic farming saw the report as validation that they were on the right track, while those who opposed organic farming saw it as a threat to their way of farming. The USDA followed up with the creation of a new position titled Organic Resources Coordinator. Shortly after the report was released and the position was created, the

Reagan Administration took over, buried the report, eliminated the position, and continued to ignore or be outright hostile to organic agriculture (Heckman 2006). The report did galvanize many to promote and advance organic farming in the policy sphere. Congressmen Jim Weaver of Oregon and Patrick Leahy of Vermont introduced organic farming bills in the House and Senate in 1982, which would develop pilot research on the effectiveness of organic farming. The bills did not gain the support of the USDA or pass into legislation (Youngberg and DeMuth 2013). For the rest of the 1980s, policy efforts refrained from using the term "organic" as it was too politically polarizing; instead, terms such as "sustainable" were used in programs and policies that drew from organic production practices.

The Organic Foods Production Association of North America (OFPANA) was created in 1985 to try to create a single national organic standard. This private sector attempt at creating a national certification failed, but the organization did go on to become a thriving association called the Organic Trade Association (Obach 2015). The first attempts at regulating the organic industry came at the state level. New York State legislated the creation of an organic advisory board in 1976. Then Connecticut, Maine, and California passed regulations for labeling organic in 1979, but the rules did not contain any provisions for oversight or enforcement (Mosier and Thilmany 2016). Within 10 years, nearly 30 more states had passed some form of organic regulations, but only four states attempted any form of certification. Washington state was the first to pass legislation regulating organic certification in 1988. By the late 1980s, there were a number of state certification programs, private certifying bodies, and farmer organizations offering certification, each with their own definition of organic and related organic standards (Youngberg and DeMuth 2013).

Processors and wholesalers were frustrated with the mishmash of certification and marketing strategies, and there were serious concerns about fraud. Those making products with multiple ingredients, especially when they came from different regions, could not get their products certified as organic or sell across state lines with different organic regulations. Consumers were confused by the numerous labels and certifiers, and that prompted many to call for a national regulation to create confidence and growth in the sector. While many people were thrilled that organic agriculture had reached such a significant milestone, many others within the organic community were opposed to a national standard because they feared it would lead to an erosion of the foundational values of organic and the loss of control to the conventional food industry. Despite their concerns, many within the organic movement supported it even though they knew it would lead to a weaker commitment to organic values, because they believed growth in the industry was worth it. A number of long-time proponents of organic farming believed the benefits of increased organic production and food were worth the shift to an industrialized organic food sector (Youngberg and DeMuth 2013). Others were worried that the cost and paperwork would overwhelm small farmers and benefit large-scale operations. Despite the concerns many had, the need for a national organic standard was becoming increasingly clear as the demand for organic food continued to grow.

Passage of the Organic Foods Production Act

In 1988, Kathleen Merrigan was a senior staff member of the U.S. Senate Committee on Agriculture, Nutrition and Forestry, led by Senator Patrick Leahy. She began work on drafting a national organic standard, reaching out to consumer and environmental organizations to support the creation of the legislation. At the time, organic farmers were not in a position to engage with policy at the federal level. There were no national organizations ready to represent organic farmers or lobby on their behalf, and many were doubtful that any meaningful policy could come from a USDA that was historically very hostile to organic farming. Despite their hesitation and lack of lobbying abilities, Merrigan met with a delegation and attempted to include their feedback in the bill she was preparing (Obach 2015).

In 1990, Senator Leahy again introduced federal legislation for organic farming as part of the Farm Bill. The bill was meant to establish an organic label that met national standards set by the USDA as a way to ensure consumers were protected against fraud and to ease interstate commerce and the growing organic processing and wholesaling industry. Leahy was also interested in providing an incentive that would encourage more farmers to convert to organic. This time the bill received support beyond the organic movement, including from the National Association of State Departments of Agriculture, the American Farm Bureau Federation, and the Center for Science in the Public Interest (Rawson 2003). The Organic Food Production Act (OFPA) was successfully passed as part of the 1990 Farm Bill with significant bipartisan support. Part of the reason for that support was the way the OFPA was envisioned and subsequently implemented. Congress stipulated the formation of a National Organic Program to be administered by the USDA's Agricultural Marketing Services (AMS) department. The USDA considers the OFPA a marketing and labeling designation and does not allow for the term "organic" to indicate any benefit over conventionally produced food.

Congress had required the costs of running the NOP to be covered by fees collected through accreditation and certification, but on several occasions, it allocated funds to be made available for cost-sharing programs for small-scale farmers and processors and another allocation for developing value-added products. The act included all organically produced food products and some health and beauty products, but it did not cover fibers such as cotton. The regulations stipulated that any farmer or business that handles more than $5000 annually of organic food must be certified by a USDA-accredited certifier in order to use the USDA organic label. No one is allowed to use the term "organic" unless they have met all requirements and been inspected by a certifying agent. Any product labeled with a USDA organic seal has to be made with at least 95 percent organic ingredients. Products made with at least 70 percent organic ingredients can say "made with organic," but cannot use the label. Private certification labels can be used in addition, but not instead of, the USDA seal (USDA AMS 2018). These regulations were created with the intention of setting a minimum standard for organic production and include provisions that allow for states to adopt stricter rules as long as they receive approval from the NOP.

Beyond developing the national standards for organic agriculture products, the NOP manages and accredits a network of third-party certifiers who do the actual certifying of farms, processors, distributors, and retailers involved in organic production (Barker 2016). The program is also responsible for enforcement, investigating fraud, and facilitating international equivalency programs. The NOP facilitates the creation of the National List of Allowed and Prohibited Substances (NLAPS), which establishes the basis of any inputs allowed in certified organic food production. The USDA does not have the resources to review all the substances that could possibly be considered, so it relies on third-party reviews. The Organic Materials Review Institute (OMRI) is an independent nonprofit that reviews materials and input products for compliance with the NOP rules (Barker 2016).

Included as part of the OFPA was the creation of a National Organic Standards Board (NOSB) to give representation to a diverse set of stakeholders in developing the standards and reviewing the list of allowable and prohibited substances or products. The NOSB was tasked with providing recommendations to the USDA on the policies and standards for certification. Congress originally set October of 1993 as the launch of the NOP, but a long series of delays occurred, setting back the launch by nearly a decade. Two years after the passage of the OFPA, the NOSB finally received funding to begin the process of creating recommendations. The board submitted their recommendations to the USDA in 1995, and it took another two years for the USDA to release the first draft rule for review. These first draft rules published in the Federal Register for public comment differed from the recommendations put forth by the NOSB in several significant ways. These major issues became known as the "Big Three," including allowing the use of GMOs, irradiation, and bio-sludge or sewage sludge. There was significant backlash from the organic community, especially from organic consumers. The USDA received an unprecedented 275,603 public comments, more than any other proposed rule in government history, and it sparked the creation of the Organic Consumers Association. The controversy prompted the USDA to withdraw the rules and rewrite them (Clark 2015). The final rules were entered into the Federal Register in 2000, and the National Organic Program (NOP) began labeling products with the USDA organic seal in 2002, more than a decade after the OFPA was passed.

The difficulty in creating consensus around the principles and practices in the organic community lasted long after the passage of the OFPA and final rules. Situating organic agriculture as a marketing and labeling program within the USDA meant that the NOP had the challenge of creating strict standards without implying that the use of chemicals, synthetics, and GMOs in agriculture was a negative attribute of conventional agriculture. The pace of growth of organic agriculture only increased the animosity and divisions within the organic community, and it has created ongoing conflict around many rules and regulations that the NOSB has considered and made recommendations on over the years. In fact, the NOP faced challenges as soon as the program launched.

In 2002, a blueberry farmer and organic inspector, named Arthur Harvey, filed a lawsuit against then Secretary of Agriculture Anne Veneman, challenging several aspects of the organic rule as not complying with the OFPA. He won three out of nine counts on appeal in 2004 with the support of a coalition of nonprofit groups. The lawsuit caused upheaval in the organic community with supporters on both sides arguing that there were many unintended consequences of the ruling that would impact farmers. The main tenets of the Harvey argument were that several elements of the rules were not in line with the principles of the OFPA including: the rule exempting from certification any nonorganic products not commercially available, the provision allowing synthetic substances in processing, and the rule allowing 20 percent conventional feed to be included during a dairy herd's conversion to organic (Viña 2006). Those in favor of the rules as they stood argued that the organic industry would not be able to grow without the use of synthetic substances in processing and that forbidding nonorganic products and feed would impose unfair financial difficulties on farmers. Following the partially successful appeal, Congress passed an amendment to the OFPA rendering the decision invalid. This has continued to rankle organic farmers who believe the NOP has strayed too far from the original principles of organic.

Federal Support of Organic since Passage of OFPA

Government support for organic agriculture has continued to increase at the federal and state levels. Despite the growing support, it took a long time for Congress and the USDA to develop programs that serve organic farmers well. In 2009, Kathleen Merrigan, the original author of the OFPA, was appointed Deputy Secretary of Agriculture under the Obama administration. Under her oversight, it was determined the NOP had grown enough to warrant its own Deputy Administrator (Clark 2015). Miles McEvoy, a 20-year veteran of the Washington state organic program, was the first to be hired in this role. Together, they made a concerted effort to ensure that organic production was integrated into every part of the USDA and that all USDA staff had a minimum knowledge of organic farming. Further training was made available for field staff, and more effort was given to issues of fraud, export markets, and organic farmer supports.

Many countries around the world have their own organic regulations and standards. For a long time, organic farmers needed to make their own arrangements to ensure their products met the standards at each location they were shipping their products to. In 2009, the United States arranged its first certification equivalency program with Canada, making the export of organic products much easier. Other locations were added including the EU, Japan, South Korea, Switzerland, and Mexico. Additional arrangements are underway with more countries.

There are now a number of programs in existence that support organic farmers with transition, risk management, and trade opportunities. One major program that was created early on and has continued throughout the years is a cost-share program to assist farmers in gaining certification. Getting certified includes application fees, inspection fees, and sometimes laboratory testing fees. The program covers up to 75 percent of the costs. Cost-share programs are especially important for farmers as they move through the three-year transition phase of certification. During that time, they will not be certified organic and will not receive a price premium. The cost-share program helps offset the cost of certification and the typically lower yields they produce during the first few years of organic production.

In 2002, the U.S. Farm Bill included over $32 million for organic agriculture, including a cost-share program for farmers transitioning to certified organic and research on organic production. It also included $15 million over three years to launch an Organic Research and Extension Initiative (OREI). This still accounted for a very small portion of the overall USDA research budget, but it was a start. Then in 2004, the USDA's Cooperative State Research, Education, and Extension Service (CSREES) allocated another $4.7 million to fund the OREI and a new Organic Transitions Program. In the 2008 Farm Bill, the funding increased to $112 million over five years.

The 2014 Farm Bill brought another expansion of funding for the organic industry. The certification cost-share program was expanded along with the funds for organic research. It also exempted organic farmers from having to pay checkoff fees (fees to cover various marketing and research programs mainly associated with conventional producer groups). The bill also created an option for a stand-alone organic checkoff program. In addition, the bill requested that crop insurance programs be improved for organic farmers.

For many years, organic farmers were unable to take advantage of the federal crop insurance program because it did not extend coverage to include the price premium that organic farmers receive. That has now changed, and farmers are able to insure their commodity crops. However, there is still no crop insurance that covers the diversity of crops that organic farmers grow or those who do not grow commodity crops. In 2016, the USDA's Risk Management Agency eased another burden from organic farmers by creating a new crop insurance program that covers diversified farms. Historically, organic farmers were unable to receive crop insurance due to their management practices that included green manures, crop rotation, and more diversification than most conventional farmers.

The 2018 Farm Bill included some big successes for the organic farming community. It included a provision for $50 million in annual funding for the Organic Agriculture Research and Extension Initiative to begin by 2023. This funding commitment makes the program a mandatory program in the USDA budget, which lends some stability to organic farmers and researchers. The program had been funded at only $20 million a year for more than a decade. Crop insurance coverage was improved again by clarifying that organic prices were to be used in determining payouts. Additional supports were added to transitioning farmers, and the export market program was mandated to include organic commodities. Another big addition was funding for data collection and enforcing organic standards on imported foods.

Data collection is a valuable element of government support as it lends legitimacy to organic production practices and gives government staff, policymakers, and advocates the knowledge they need to make good decisions about where to allocate funds. The organic sector has been, and still is, underrepresented in the data collection and reporting done by many USDA departments. Much of the available data about organic food and beverage sales, research, consumer demand, and other issues has been collected by nongovernmental organizations and therefore is not freely available to the public.

Starting in 1992, the USDA ERS gathered data from certifiers and used that data to calculate the amount of organic acreage and production. Then in 2008, the Farm Bill included provisions for the National Agricultural Statistics Service (NASS) to conduct national surveys of organic agriculture. In 2008 and 2015, the NASS included an organic section in the general agricultural census, which captured a small amount of information specific to the organic sector. In addition, independent certified organic surveys were designed and conducted in 2011, 2015, 2016, and 2019 to gather more comprehensive data on the organic industry.

The Growth of an Organic Sector

Organic food has been one of the fastest growing segments of the food industry since certification was introduced in 2002. Even before certification, the growth was substantial. In 1997, U.S. organic sales were estimated to be $3.6 billion (Dimitri and Oberholtzer 2009). Beginning in the late 1990s, retail sales of organic products began increasing by about 20 percent a year, and that trend continued for over 20 years. The growth slowed during the recession of 2008 to about 5 percent and has fluctuated between about 6 and 12 percent growth in the years since then (USDA ERS 2018).

By 2016, the total sales of organic products reached over $47 billion (Willer and Lernoud 2018; National Institute of Food and Agriculture 2018), which was over 4 percent of the total food sales in the United States (USDA ERS 2018). Compared to the total food sales, organic sales are not that large, but the percentage of growth is huge, and that has attracted the attention of many supermarkets and conventional food suppliers. Fresh fruits and vegetables are the most sought after organic foods followed by dairy, and they have been since retailers began offering organic food. In 2019, the organic market in the United States was worth $55.1 billion with $50.1 billion in food sales, which is just under 6 percent of all U.S. food sales (OTA 2019). That makes the United States the largest organic market in the world (Willer and Lernoud 2018), although North America has only 6 percent of the world's organic agricultural land. Fruits and vegetables were the most popular organic products making up 36 percent of all organic food sales and almost 15 percent of all fruits and vegetables sold in the United States (Gelski 2019). The onset of the COVID-19 pandemic dramatically increased consumer demand for organic food with sales increasing 20-50 percent in the first half of 2020 (OTA 2020).

Studies of organic consumers in the past showed that most purchasers were females who had higher incomes and levels of education. However, those demographics are changing drastically and can no longer be used to describe the typical organic consumer (Aschemann et al. 2007). While it is true that households with limited food budgets tend to spend less on organic food, other factors are stronger indications of the willingness to pay for organics (Aschemann-Witzel and Zielke 2017). A Gallup poll in 2014 found that 45 percent of Americans sought out organic food for their households. More recent studies have found that that has increased to 85 percent of U.S. households purchasing at least some organic food (OTA 2018). That percent was even higher for the millennial population (Greene et al. 2017). The dominant motivation for purchasing organic food is health, usually because people want to avoid pesticides, synthetic additives, hormones, antibiotics, and GMOs. Many consumers also believe that organic foods are more nutritious. Other reasons consumers choose organic food are better animal welfare, environmental benefits, better taste, and support for local economies (Kim et al. 2018).

The biggest barrier to consuming organic food is the cost of organic food, but availability is also an important consideration in terms of both product range and where organic food is sold. Although the price differences between organic and conventional foods are decreasing in many cases, there are still a number of products that command a big price premium. Between 2004 and 2010, all organic products were more expensive than conventional products, although the difference between the two fluctuated (Carlson and Jaenicke 2016). The biggest price differences are generally for milk and other dairy products and eggs, followed by salad mix. Even when there aren't any major price differences, consumers anticipate a premium and shy away from purchases of organic food. Thus, organic retains the reputation of being elitist and expensive (Aschemann-Witzel and Zielke 2017).

In 2016, the United States imported organic food worth $1.65 billion with bananas, coffee, and olive oil topping the retail food list. Organic feed grain was also imported in large quantities. Exports accounted for $40 million in 2016 with apples, grapes, and lettuce shipped primarily to Canada and Mexico. A major struggle in the growth of the organic food sector is that demand has continued to outpace supply. This is especially true for organic meat as it takes more resources to raise livestock than vegetables and fruit. Grain production is the largest restriction for livestock producers who cannot source enough organic feed grain (Greene et al. 2009).

Despite this growing demand, organic still accounts for less than 2 percent of total agricultural land around the world (Willer and Lernoud 2018). Organic market growth expanded significantly starting in the 1980s, and certified organic cropland expanded as well. Certified organic farming systems comprised 1.35 million acres of cropland and pasture in 49 states by 1997. The amount of organic cropland in the United States more than doubled between 1992 and 1997, but the increase in the number of certified growers was much less. This can be attributed to farmers expanding operations and new larger-scale operations becoming certified. Organic acreage again doubled between 1997 and 2005 (Dimitri and Oberholtzer 2009).

Twenty years later, there are now over 26,000 certified organic farms in the United States (OTA 2018). A report by Mercaris showed 3.3 million acres of certified organic commodities were grown in the United States in 2019. That is a 14 percent increase in the number of growers in operation, double the increase from the previous year. The total certified organic acreage in 2018 was 8.3 million acres with the majority in California, followed by Montana, North Dakota, and Minnesota. While most of the large-scale fruit and vegetable production occurs in California, making it the top producer in that category, the Northeast has the highest number of certified organic farmers, many of them small scale (Greene et al. 2017). Historically, the South has had the fewest organic operations, although there has been substantial growth in recent years. Regional difference can be attributed to the suitability of certain types of production for each area, access to urban markets versus commodity markets, and historical precedent.

Selling in the Local Food Market

In early days, most organic large-scale commodity farmers merely sold their crops and livestock in the conventional markets because that was their only option. Despite the lack of options, many organic farmers wanted to connect with consumers who shared their same values, and they often found that they could generate a higher profit if they sold their products directly to the consumers through farmers' markets, roadside shops, or community-supported agriculture (CSA). Through the 1990s, approximately one-third of organic producers directly marketed their produce to restaurants, grocery stores, and through direct contact with consumers (Buck et al. 1997). As consumer demand for local organic products increased, the growth of farmers' markets followed suit. In 1994, there were 1755 active farmers' markets in the country, and by 2019, the number of markets had increased to 8771 (USDA AMS 2020). Alongside the growth of farmers' markets and other direct sales venues, there has been an increase in local and regional food networks that provide marketing assistance, local branding, and directories of farms often giving importance to organic farms. Some of this growth in high-value local food production has been due to demand from restaurants.

Meanwhile, several high-profile chefs had started introducing organic food to the gourmet food community. Nora Pouillon started an organic restaurant in Washington, DC, in 1979. She based her restaurant's menu and approach to food as one having a connection with local farmers and the seasons of farming. Nora managed to move beyond the expectations that organic food was just hippie vegetarian food and showcase gourmet meals. Her influence led other chefs to seek out direct relationships to organic farmers who could deliver high-quality ingredients. These relationships established an important niche market for small-scale organic farmers. When organic certification became widely known, Nora took on the task of adapting the standards to the restaurant industry. Her restaurant was the first and has remained one of the few restaurants to maintain organic certification. Nora also founded a farmer's market that brought even more market opportunities to the region's organic farmers. Across the country, another chef was creating the same kind of buzz in California. Alice Waters started Chez Panisse in 1971 with a group of friends. The legendary restaurant has always made local organic ingredients a core principle of the operation. Their model of sourcing directly from local suppliers who used organic production methods inspired many others to try the same approach.

Many consumers who are interested in buying organic food are also interested in purchasing from local producers. Consumers who buy local organic food have many of the same motivations as those who buy local, and in fact, at least one study has shown that a third of organic consumers felt buying local was more important than buying organic. Many consumers prefer to buy organic locally because they feel more confident in local sources, the food is fresher and tastes better, and they are supporting their local economy and community (Oberholtzer et al. 2014). The roots of organic supply chains are local and short. As organic demand began to outgrow that phase, many considered it to be similar to conventional agriculture with long supply chains and no personal connections; many consumers and retailers started to find ways to bridge the local and organic markets. Many people think of local food being sold just at farmers' markets, CSAs, and farm stands, but once the USDA began certifying organic food, there were an increasing number of local retailers aggregating local organic food and selling it to the public. There is often a tension between local and organic food with consumers prioritizing local over organic in many instances. Even larger stores are trying to emphasize local when they have the option. Retailers face challenges in responding to competing consumer demands for not only local food but also out-of-season or imported organic foods. Smaller stores with a long history of buying organic are more likely to buy directly from organic farmers, especially those that also focus on diverse products from local sources (Oberholtzer et al. 2014). Larger chain stores have more difficulty in sourcing directly from farmers due to the need for much higher volumes of product and a year-round supply.

Significant market growth resulted in new supply chains being created to serve the organic sector, a growth in organic imports, and the involvement of the conventional food sector (Jaenicke et al. 2011). Organic farmers started experimenting with new marketing outlets and supply chains. Some went the route of creating or connecting with the local food movement, while others scaled up their businesses to compete in the conventional sector.

Scaling Up and Building Organic Supply Chains

Decades after the launch of the NOP, over 90 percent of organic sales occur in conventional supermarkets and natural food stores with only a small percent occurring in farmers' markets and direct sales. This is a distinct shift from the ways and locations in which organic food was historically sold. Throughout the 1990s, most organic products were sold at natural food stores, and only about 7 percent was found in supermarkets. This has changed dramatically, with the year 2000 marking the first time that more organic food was purchased at supermarkets than at any other venue. Consumers spent $7.8 billion on organic food and purchased 49 percent of that in supermarkets. By the early 2000s, 73 percent of all U.S. supermarkets sold organic food (Dimitri and Greene 2002).

In 1980, the first Whole Foods Market opened in Austin, Texas, and several natural food distributors increased their organic food options. Earthbound Farms, founded in California by Drew and Myra Goodman, started marketing their spring salad mix to high-end restaurants. Then in the late 1980s, food scares including food contaminated with pesticides that caused widespread illness started occurring. In 1989, a CBS television program called *60 Minutes* did an exposé on Alar, a plant growth regulator used primarily in apple orchards (Gordon 2011). These food scares led to a rapid increase in consumer demand for organics. As organic production became more acceptable and profitable, corporations started getting involved, mostly by integrating and expanding existing operations. By the early 2000s, organic growth was so fast paced that demand regularly outpaced supply. Retailers were under pressure to meet demand and keep price premiums at a minimum. Importing supplies of organic products that do not grow in the United States or out-of-season products has increased as well. Retailers have also started using imports to bolster supply when domestic options run short (Jaenicke et al. 2011).

The organic market has undergone several major rapid structural changes as growth of organic sales increased. First, competition for shelf space has increased and so has demand for a broader range of organic products. Currently, organic products can be found in 75 percent of all grocery categories (OTA 2018). This puts considerable strain on the small processors and producers who started the organic market. They often can't compete with larger companies who can make use of economies of scale to lower the prices and provide more flexibility to retailers and suppliers. Second, small organic companies have grown to meet demand and, in the process, have often merged with other companies concentrating the market. As the companies grow, their ability to remain committed to their original values is challenged. Third, the conventional supply chains have entered the organic market, either by starting their own line of organic or organic versions of their brands or, more often, by buying organic companies and maintaining the brand (Aschemann et al. 2007; Howard 2009). In many cases, when a large company launches an organic line of products, they in fact contract another organic company to produce the product under their label (Howard 2009). The same is true when retailers introduce their own branded line of products. Thus, there is an illusion of choice present in the grocery store, but most products are processed, packaged, and distributed by only a handful of companies.

Large chains have been taking over as the main retailers of organic foods in the United States. As early as 2006, Walmart had become the largest retailer of organic milk. By 2015, Costco surpassed Whole Foods as the largest retailer of organic food with over $4 billion in sales. In 2017, Whole Foods Market was bought by Amazon, and in 2018, Walmart introduced an organic pantry line that has no price premium. The agri-business control of organic food is not confined to retailers. Large transnational food companies such as Heinz and General Mills are using their vertical integration to leverage economies of scale and navigate the certification process, especially for imported foods from the global south (Raynolds 2004). Most of the mergers and acquisitions of organic food businesses occurred between 1997 and 2000 when the USDA made it clear that a standard would be forthcoming soon. Consumers are often unaware of large corporate involvement in the organic market because the consolidation or expansion is hidden behind product branding that does not reveal the ownership of a particular product (Howard 2009).

Along with the growing interest and demand in organic agriculture and the resulting shift from direct sales to extended supply chains, organic food purchasing went from being based on relationships to being based on certification systems. This challenged the organic community as many farmers had chosen organic because of ideological factors such as a connection to nature, environmental and health concerns, and independence rather than being motivated by profits. Those who were not connected to a large like-minded community of consumers or were too far from processing, wholesale, and retail outlets had to sell into the conventional food channels forgoing the price premium that organic could bring in. Despite consumer interest in all that organic offers, many still don't fully understand what exactly organic means. Organic labels, and the standards behind them, have become increasingly important as organic has gone into mainstream retail chains and the number of organic products and brands has increased. People wanting to buy and support organic agriculture often have no other way to verify that their values are being represented in the food they purchase. For some, this was motivation to pursue a regional or national certification system and grow an organic food industry; for others, organic farming was a labor of love, and they worried that certification would mimic the conventional sector and destroy the philosophical tenets they believed in.

Some of the early successful organic farmers, such as Paul Keene of Walnut Acres, Gene Kahn of Cascadian Farms, and Gary Hirshberg and Samuel Kaymen of Stonyfield Organic, built processing facilities and started manufacturing processed organic food. All of those early farmer-owned businesses are now owned by large corporations. Farmer-owned cooperatives offered one way of building local markets and competing with large corporations, while keeping the profits in the farmers' hands. One example is the Coulee Region Organic Produce Pool Cooperative, which was the largest organic farmers' cooperative in the United States. In the early 2000s, the cooperative had over 500 farm families involved and was the only large organic cooperative to be owned and operated completely by farmers. While alternative forms of ownership and supply chains do exist, the vast majority of organic food is farmed, processed, distributed, and purchased in the same model as conventional agriculture.

Challenges Come with Growth

Introduction of a national organic standard changed the organic community. For some farmers, the new standard provided an opportunity, and for others, it was a loss. Not all farmers made the decision to become certified once the program became active. At a practical level, any farmer who wanted to label their food as organic and made more than $5000 in sales did need to become certified.

As the market grows to meet the demand, more farmers are needed to grow organic food. Despite the possibility of receiving a premium price, especially for organic grain crops, not enough farmers are converting. Some of the barriers to conversion are related to the specialized knowledge that organic farming requires. Many conventional grain farmers are comfortable with a corn/soybean rotation, but organic farming requires additional crop rotation and other management practices such as including crops that are not sold but put back into the soil. Some of the other major challenges that organic farmers face include pesticide and GMO drift, bee colony collapse, sourcing organic seeds and other inputs, and concentration, especially in the dairy sector. New farmers have trouble gaining access to land and capital that is needed to start farming, especially on a larger scale required for grain production.

Another challenge facing organic farmers comes from the success of the organic industry. Large corporations were not the only ones to see opportunity in organic agriculture. Conventional farmers have seen many benefits of organic farming from their neighbors or at the markets and have transitioned to organic farming in increasing numbers. Historically, there have been large shifts to organic production of grains when prices for organic commodities peak, and then there is a drop in organic production when the prices start to fall again. This mirrors the rise and fall of commodity prices in the conventional sector, but this can have an impact on the number of farmers that choose to transition to organic or even remain certified. The effects of agribusiness engagement with organics also drove a large transition to organic farming to meet the growing demand for organic processed foods. This can often lead to small values-committed organic producers to be squeezed out of the market (Guthman 2004), something especially true in the dairy sector. Access to larger markets can also represent the types of farmers that choose to be certified. Farmers who have production contracts have access to regional distribution centers or sell high-value crops that tend to be certified organic (Uematsu and Mishra 2012).

In addition to these specific challenges faced by organic farmers, it is important to acknowledge that organic farmers operate in a system that has been designed for conventional agriculture on all fronts. Conventional agriculture has benefited from decades and billions of dollars of research, business development, and policy efforts. Organic agriculture has only received a small fraction of attention from researchers and from research dollars. U.S. policies have been structured to support conventional agriculture, and despite new interest in organic agriculture support, the majority of information, support programs, and subsidy dollars benefit conventional farmers, not organic farmers. Industry designed for the conventional agriculture sector has not really adapted to organic, but rather the organic sector has "conventionalized" to fit in (Guthman 2004). Despite this lack of support, the organic sector has grown and shows huge potential to address many of the drawbacks of conventional agriculture.

References

Aschemann, J., U. Hamm, S. Naspetti, and R. Zanoli. 2007. "The Organic Market." In *Organic Farming: An International History*. Oxford: CABI.

Aschemann-Witzel, Jessica, and Stephan Zielke. 2017. "Can't Buy Me Green? A Review of Consumer Perceptions of and Behavior Toward the Price of Organic Food." *Journal of Consumer Affairs* 51 (1): 211–51.

Barker, Allen V. 2016. *Science and Technology of Organic Farming*. Boca Raton, FL: CRC Press.

Barton, Gregory A. 2018. *The Global History of Organic Farming*. Oxford: Oxford University Press.

Buck, Daniel, Christina Getz, and Julie Guthman. 1997. "From Farm to Table: The Organic Vegetable Commodity Chain of Northern California." *Sociologia Ruralis* 37 (1): 3–20.

Carlson, Andrea, and Edward Jaenicke. 2016. "Changes in Retail Organic Price Premiums from 2004–2010." USDA: Economic Research Service.

Clark, Lisa. 2015. *Changing Politics of Organic Food in North America*. Northampton, MA: Edward Elgar Publishing Ltd.

Coleman, Elliot. 1995. *The New Organic Grower: A Master's Manual of Tools and Techniques for the Home and Market Gardener*. Vermont: Chelsea Green Publishing Company.

Conkin, Paul Keith. 2008. *A Revolution down on the Farm: The Transformation of American Agriculture since 1929*. Lexington: University Press of Kentucky.

Davis, Frederick Rowe. 2014. *Banned: A History of Pesticides and the Science of Toxicology*. New Haven, CT: Yale University Press.

Dibner, J. J., and J. D. Richards. 2005. "Antibiotic Growth Promoters in Agriculture: History and Mode of Action." *Poultry Science* 84 (4): 634–43.

Dimitri, Carolyn, and Catherine Greene. 2002. "Recent Growth Patterns in the US Organic Foods Market." Agriculture Information Bulletin 777. USDA/Economic Research Service.

Dimitri, Carolyn, and Lydia Oberholtzer. 2009. "Marketing U.S. Organic Foods: Recent Trends from Farms to

Consumers." Economic Information Bulletin 58. USDA: Economic Research Service.

Duram, Leslie A. 2005. *Good Growing: Why Organic Farming Works*. Lincoln: University of Nebraska Press.

Ellis, B., and F. Bradley, eds. 1996. *The Organic Gardener's Handbook of Natural Insect and Disease Control*. Emmaus, PA: Rodale Press, Inc.

Foster, John Bellamy, and Fred Magdoff. 1998. "Liebig, Marx, and the Depletion of Soil Fertility: Relevance for Today's Agriculture." *Monthly Review; New York*, August.

Gelski, Jeff. 2019. "Sales Increase 5.7% for U.S. Organic Foods." *Baking Business*, May 20.

Goldberger, Jessica. 2011. "Conventionalization, Civic Engagement, and the Sustainability of Organic Agriculture." *Journal of Rural Studies* 27 (1): 288–96.

Gomiero, Tiziano, David Pimentel, and Maurizio G. Paoletti. 2011. "Environmental Impact of Different Agricultural Management Practices: Conventional vs. Organic Agriculture." *Critical Reviews in Plant Sciences* 30 (1–2): 95–124.

Gordon, Wendy. 2011. "The True Alar Story." *Huffington Post*, March 30. https://www.huffingtonpost.com/wendy-gordon/the-true-alar-story_b_838974.html.

Greene, Catherine, Carolyn Dimitri, Biing-Hwan Lin, William McBride, Lydia Oberholtzer, and Travis Smith. 2009. "Emerging Issues in the U.S. Organic Industry." *United States Department of Agriculture, Economic Research Service, Economic Information Bulletin*, January.

Greene, Catherine, Gustavo Ferreira, Andrea Carlson, Byce Cooke, and Claudia Hitaj. 2017. "Growing Organic Demand Provides High-Value Opportunities for Many Types of Producers." *Amber Waves*, Jan/Feb.

Gustafson, R. H., and R. E. Bowen. 1997. "Antibiotic Use in Animal Agriculture." *Journal of Applied Microbiology* 83 (5): 531–41.

Guthman, Julie. 2004. "The Trouble with 'Organic Lite' in California: A Rejoiner to the 'Conventionalisation' Debate." *Sociologia Ruralis* 44 (3): 301–16.

Heckman, J. 2006. "A History of Organic Farming: Transitions from Sir Albert Howard's War in the Soil to USDA National Organic Program." *Renewable Agriculture and Food Systems* 21 (3): 143–50.

Howard, Albert, Sir. 1940. *An Agricultural Testament.* London; New York: Oxford University Press.

Howard, Philip. 2009. "Visualizing Consolidation in the Global Seed Industry: 1996–2008." *Sustainability* 1 (4): 1266–87.

IFOAM. 2018. "Leading Change Organically: 2017 Consolidated Annual Report of IFOAM-Organics International." IFOAM Organics International.

Ikerd, John. 2016. "Family Farms of North America." Working Paper 152. Brazil: UN FAO and UNDP.

Jaenicke, Edward, Carolyn Dimitri, and Lydia Oberholtzer. 2011. "Retailer Decisions about Organic Imports and Organic Private Labels." *American Journal of Agricultural Economics* 93 (2): 597–603.

Kim, GwanSeon, Jun Seok, and Tyler Mark. 2018. "New Market Opportunities and Consumer Heterogeneity in the U.S. Organic Food Market." *Sustainability* 10 (9): 3166.

Kirchhelle, Claas. 2018. "Pharming Animals: A Global History of Antibiotics in Food Production (1935–2017)." *Palgrave Communications* 4 (1): 96.

Leigh, G. J. 2004. *The World's Greatest Fix: A History of Nitrogen and Agriculture.* Oxford; New York: Oxford University Press.

Martin, D., and G. Gershuny. 1992. *The Rodale Book of Composting: Easy Methods for Every Gardener.* Emmaus, PA: Rodale Press, Inc.

Mosier, Samantha L., and Dawn Thilmany. 2016. "Diffusion of Food Policy in the U.S.: The Case of Organic Certification." *Food Policy* 61 (May): 80–91.

National Institute of Food and Agriculture. 2018. "Organic Agriculture Program | National Institute of Food and Agriculture." https://nifa.usda.gov/program/organic-agriculture-program.

National Organic Standards Board (NOSB). 2000. Comments to Revised Proposed Rule 17. AMS-TMP-USDA 7 CFR PART 205.

National Research Council. 2000. "Ch. 1 History and Context." In *The Future Role of Pesticides in US Agriculture*. Washington, DC: The National Academies Press.

Obach, Brian. 2015. *Organic Struggle: The Movement for Sustainable Agriculture in the United States*. Cambridge, MA: MIT Press.

Oberholtzer, Lydia, Carolyn Dimitri, and Edward C. Jaenicke. 2014. "Examining U.S. Food Retailers' Decisions to Procure Local and Organic Produce from Farmer Direct-to-Retail Supply Chains." *Journal of Food Products Marketing* 20 (4): 345–61.

Organic Trade Association (OTA). 2018. "Organic: Big Results from Small Seeds." https://ota.com/sites/default/files/indexed_files/2018_OrganicIndustryInfographic_0.pdf.

Organic Trade Association (OTA). 2019. "COVID-19 Will Shape Organic Industry in 2020 After Banner Year in 2019." https://ota.com/news/press-releases/21328.

Paull, John. 2009. "A Century of Synthetic Fertilizer: 1909–2009." *Journal of Bio-Dynamics Tasmania*, June.

Paull, John. 2011. "The Betteshanger Summer School: Missing Link Between Biodynamic Agriculture and Organic Farming." *Journal of Organic Systems* 6 (2): 14.

Rawson, Jean. 2003. "Organic Foods and the USDA National Organic Program." Report for Congress RL31595. Washington, DC: Congressional Research Service.

Raynolds, Laura T. 2004. "The Globalization of Organic Agro-Food Networks." *World Development* 32 (5): 725–43.

Rigby, D., and D. Cáceres. 2001. "Organic Farming and the Sustainability of Agricultural Systems." *Agricultural Systems* 68 (1): 21–40.

Seufert, Verena, Navin Ramankutty, and Tabea Mayerhofer. 2017. "What Is This Thing Called Organic?—How Organic Farming Is Codified in Regulations." *Food Policy* 68 (April): 10–20.

Standage, Tom. 2009. *An Edible History of Humanity.* New York: Bloomsbury Publishing USA.

Uematsu, Hiroki, and Ashok K. Mishra. 2012. "Organic Farmers or Conventional Farmers: Where's the Money?" *Ecological Economics* 78 (June): 55–62.

U.S. Department of Commerce, Bureau of the Census. 1978. *1974 Census of Agriculture: Farms: Number, Acreage, Value of Land and Buildings, Land Use, Size of Farm, Farm Debt.* Vol. 2. U.S. Department of Commerce, Bureau of the Census.

USDA, Agricultural Marketing Service. 2018. "Introduction to Organic Practices." https://www.ams.usda.gov/publications/content/introduction-organic-practices.

USDA, Agricultural Marketing Service. 2020. "Farmers Markets and Direct-to-Consumer Marketing." https://www.ams.usda.gov/services/local-regional/farmers-markets-and-direct-consumer-marketing.

USDA, Economic Research Service. 2017. "America's Diverse Family Farms: 2017 Edition." Economic Information Bulletin 185. USDA ERS.

USDA, Economic Research Service. 2018. "Organic Agriculture." https://www.ers.usda.gov/topics/natural-resources-environment/organic-agriculture/.

USDA Study Team. 1980. "Report and Recommendations on Organic Farming." National Agriculture Library Archive. https://pubs.nal.usda.gov/sites/pubs.nal.usda.gov/files/Report%20and%20Recommendations%20on%20Organic%20Agriculture_0.pdf.

van der Ploeg, R. R., W. Böhm, and M. B. Kirkham. 1999. "On the Origin of the Theory of Mineral Nutrition of Plants and the Law of the Minimum." *Soil Science Society of America Journal* 63 (5): 1055–62.

Viña, Stephen R. 2006. "Harvey v. Veneman and the National Organic Program: A Legal Analysis." Congressional Research Service.

Willer, Helga, and Julia Lernoud, eds. 2018. "The World of Organic Agriculture Statistics and Emerging Trends 2018." Research Institute of Organic Agriculture (FiBL), Frick, and IFOAM-Organics International, Bonn.

Youngberg, G., and S. P. DeMuth. 2013. "Organic Agriculture in the United States: A 30-Year Retrospective." *Renewable Agriculture and Food Systems* 28 (4): 294–328.

Introduction

The rapid growth of organic food in the past 30 years has come though innovation, tenacity, and an ever-growing demand for it. The expansion of the organic industry has made organic food widely available in every state and nearly all grocery stores to some extent. While the overall organic sector still only accounts for around 5 percent of all food sales and just over 1 percent of all land under production, the continued growth is often seen as a threat to the conventional food sector. Much of the opposition to organic is woven into the history and goals of the agribusiness industry, but it is also linked to the larger political, social, and economic climate.

The clash of ideologies and goals of organic and conventional agriculture makes their way into research, media, and public opinion. The result is a lot of confusing information that makes it hard to determine whether organic food is in fact a good idea. To make matters worse, the organic farming movement is not a homogenous group. There are as divergent views and ideologies within the organic community as there are outside it. This leads to plenty of disagreement about what is the best way forward, especially now that the organic industry is regulated by the USDA and changes require substantial lobby efforts. As the demand for organic continues to grow,

---

Organic chickens wander around a free-range barn. Minimum indoor space and access to the outdoors have been controversial issues in the organic poultry industry. (MartinBergsma/Dreamstime.com)

there are many challenges facing organic farmers and the overall organic sector.

This chapter addresses the most common questions that arise about the ability of organic to meet consumer expectations and then explores some of the biggest challenges facing the organic industry at this time. The first half of the chapter is set up as a series of questions to answer, and it covers the biggest arguments against organic food and farming. The second half of the chapter addresses the main barriers to growth in the organic sector. Finally, the chapter closes with some ideas for bolstering the organic industry and preparing it for another 30 years of growth.

Questioning the Efficacy of Organic Food

Ever since farmers started experimenting with organic agriculture practices, they have received pushback, ridicule, and sometimes downright hostility. The earliest pioneers of organic farming were often considered to be quacks, especially as some of them had questionable beliefs about health or spirituality. J.I. Rodale, generally considered the father of the U.S. organic movement, was profiled in the *New York Times Magazine* in an article published on June 1971 called "Guru of the Organic Food Cult." It was not long after that that the Secretary of Agriculture, Earl Butz, was quoted as saying, "Before we go back to organic agriculture, somebody is going to have to decide what 50 million people we are going to let starve." The more popular organic farming became, the more resistance they got from the established agriculture community. The growing organic movement was seen as a crowd of hippies and back-to-the-landers not at all serious about business growth or science. Despite these impressions, organic food and farming grew into the multibillion-dollar economy that it is today, and its portrayal in the media has shifted to include a broad range of coverage and perspectives.

In fact, a review of 618 news stories relating to organic food and farming from 1999 to 2004 showed that 41 percent were neutral in tone and as many as 36 percent were positive (Cahill et al. 2010). This shift in coverage is directly related to consumers' perceptions of organic food. A number of studies conducted in Australia found that media coverage of issues related to pesticides, GMOs, and the environment may influence consumers' opinions and spending on organics (Lockie et al. 2002, Sloan 2002).

An in-depth look at the linkages between organic agriculture and food safety, human health, and the environment in North American media suggested that coverage of food safety scares and GMO technology tended to include references to organic farming methods (Cahill et al. 2010). Organic farming was consistently being portrayed as a positive alternative to consumers' concerns about significant food safety issues and environmental concerns, although that is certainly not always the case. In 2004, the science journal *Nature* published a couple of articles on organic agriculture that addressed many of the common questions and concerns that people have about organic farming. The articles presented opinions from a number of academics that basically said there wasn't enough definitive evidence on either side of the argument, but the overall slant of the articles seemed to suggest that the argument for organic was overblown.

A common argument against organics is that all farmers are concerned about soil health, not just organic farmers. In addition, many practices that were previously used primarily by organic farmers are becoming widespread practices for all farmers. The theory goes that organic is a marketing gimmick designed to garner a price premium, and not really all that different from conventional farming. Layered on top of that is the argument that organic farmers will not be able to meet the global food demands, especially as it requires much more land and labor. There are those who see organic farmers as being closed off to technological and scientific advances and those who see conventional farmers as pawns in a chemical industrial system that doesn't benefit anyone but large corporations.

Organic Food and Farming in the Media: Science or Lobby Tactics?

There are many news articles that question the validity of organic food and farming. Often these headlines are sparked by the release of a new study or academic article. For example, in 2012, Stanford University researchers published an article in the *Annals of Internal Medicine*, declaring that organic fruit and vegetables were no more nutritious than conventional ones (Smith-Spangler et al. 2012). The article sparked a series of headlines in the *Washington Post*, *CBS News*, and *New York Times*, questioning the validity of organic food (Philpott 2012). In 2014, there was another round of articles along the same theme, that organic is all hype and no better than conventional food (Hamerschlag 2014).

In both cases, proponents of organic wrote responses that refuted these claims and offered studies that suggested that the articles overlooked the health implications of pesticide exposure. With these opposing viewpoints, both drawing on scientific studies to back up their claims, how can consumers make informed choices? In some cases, there is legitimate evidence on both sides that makes it difficult to come to a definitive answer. Other times, the origins of an article or study can be traced to some dubious sources that are clearly trying to influence the public to further their own cause.

The organization, Friends of the Earth, investigated the sources used in many of the news media articles that portray organic in a negative light. They found that the main sources were in fact from one organization called Academics Review, which, it turns out, has ties to large agri-business, food, and oil companies (Hamerschlag 2014; Malkan 2017). Then in 2015, the *New York Times* revealed that both Monsanto and organic lobby groups had recruited academics to support their claims on both sides of the GMO debate and supported them with research dollars in exchange for lobbying. Additionally, there was evidence that Monsanto even went as far as to draft material for a website called GMO Answers (Lipton 2015). Numerous articles that attack organic agriculture in recent years can be traced back to a man named Henry Miller, a scientist who was found to be submitting articles ghostwritten by Monsanto to *Forbes* magazine (Ruskin 2017). Despite the exposé, several media outlets have continued to publish content under his name, including *Newsweek*.

So, what does the scientific community say about these issues? It can be very difficult to determine whether organic farming and food is good or not; however, there are trends in the research, and if enough research starts showing the same results over and over again, then the answers become a little clearer. It can also be important to look at who is doing the research, how they are funded, and if they are examining the issue within the right context.

The following sections will examine several of the most common arguments for or against organic agriculture and examine the most recent evidence available. The themes explored will include viability of the organic agriculture sector, health and nutrition of organic food, environmental benefits or drawbacks of organic farming, the impact of organic agriculture on the climate crisis, and finally the role of organic in social justice issues.

Is Organic Food Production Viable?

One of the overriding concerns people have is organic agriculture's potential to produce enough food to feed the world and remain economically viable. Studies on organic and conventional yields and profitability are attempts to address these concerns, and they show a wide range of results (Meemken and Qaim 2018). Studies have shown that organic yields are on average about 20 percent lower than conventional yields on a global scale (de Ponti et al. 2012; Ponisio et al. 2015; Seufert and Ramankutty 2017). The global results though vary considerably depending on a wide range of factors. Comparison studies conducted on long-term research sites show a different conclusion.

Growing food organically takes time to both build soil to productive levels and also for a farmer to gain skills in managing a different style of production (Delate et al. 2015). Properly developed crop rotations and management systems results in better soil fertility in the long term, which results in yields that are on par or even better than conventional rotations in comparison studies (Delate et al. 2015; Schrama et al. 2018). Another study found that organic systems can actually reduce the yield gap when multiple crop rotations were incorporated into a farm (Ponisio et al. 2015).

The argument that lower organic yields will make it impossible to grow organic food for a worldwide population does not take into account the time it takes to build soil fertility and knowledge, but once those have been achieved, there is no reason organic cannot maintain supply. In a study published in 2006, a group of researchers examined 293 examples of yield comparisons on a global scale and found that global food supply could be met on a per capita basis with current organic production output and without having to increase the agriculture land base. Furthermore, they found that using leguminous cover crops supplied sufficient nitrogen to replace all synthetic fertilizers (Badgley et al. 2007).

Studies have consistently shown organic farms to be more profitable than conventional farms when receiving a price premium. An analysis of global organic and conventional production systems showed that organic agriculture usually received a price premium around 30 percent, but that they only needed to receive 5–7 percent to break even with conventional profits (Crowder and Reganold 2015). Other studies suggest that while gross sales are higher for conventional crops, net income is about the same when all costs are considered (Uematsu and Mishra 2012). Generally, the overall management costs were similar although the breakdown is different with organic agriculture having higher labor costs but lower input costs. Income is often lower than that in conventional farms because of lower yields, but that is usually more than offset by the premium prices (Crowder and Reganold 2015; Seufert and Ramankutty 2017). When producing grains, the economic returns improve if perennial forage is included in the rotation (White et al. 2019). Organic farmers have not always had an easy time making ends meet as there are significant challenges that organic farmers face. Most of the agriculture industry is set up to support the conventional sector, and that can hinder factors that impact yields and profits. Organic farms require more labor, especially on fruit and vegetable farms (Gomiero et al. 2011). Organic farmers often take on more risk as there were historically few options for insurance or traditional farm loans. New studies are adopting the whole farm approach to determine profitability on farms, especially on organic farms (Delbridge 2014). This approach can provide a better overall picture of the successes and challenges in organic farming because it takes into account the full context of the farm instead of comparing just one set of crops.

One group of researchers have suggested subsidizing organic food production by providing organic farmers with a payment on organic fruit or vegetable production to make it more accessible and affordable (Nelson et al. 2019). This approach could benefit both farmers and consumers, but it would take a considerable change in the political climate.

Is Organic Food Healthier and Safer to Consume?

The debate about the nutritional or health benefits of eating organic food has been ongoing for several decades now, and while progress has been made, there are no definitive answers yet. There are a number of established nutritional differences between organic and conventional foods, but there is no proof yet that these differences translate into changes in health (Barański et al. 2017). There are very few studies that look at the health implications of eating organic food, and essentially none are long term. Most studies are based on self-reported consumption and therefore have a tendency to contain some measurement errors. The few studies that are more controlled draw on a very small sample size for a limited time, which in turn, limits the statistical significance and understanding of long-term impacts. Other challenges come from the fact that people who choose organic foods are also more likely to have generally healthier eating habits and other attributes that lower incidence of disease (Mie et al. 2017).

Organic foods have higher levels of antioxidants, which can reduce the likelihood of developing some diseases (Barański et al. 2017). Organic dairy and meat have higher levels of Omega-3 fatty acids, while organic dairy has higher levels of conjugated linoleic acid, tocopherols, and iron concentrations, which are all nutritionally desirable. Organic milk was shown to have lower amounts of selenium and iodine, which could be concerning in places where iodine is not readily found in the diet. Conventional foods tend to have higher levels of toxic heavy metals such as, cadmium, pesticides, and compounds such as nitrite and nitrate (Barański et al. 2017). Likewise, conventional meat has higher levels of saturated fats, which are thought to contribute to cardiovascular disease.

The main source of exposure to pesticides for the general population is through food consumption, especially fruits and vegetables. Farmworkers and those living in rural areas have high external exposure as well. For the most part, studies that looked at pesticide intake showed those on an organic diet had much lower exposure to pesticides (Mie et al. 2017). A number of studies that measured concentrations of pesticides in urine after a week of having only organic food showed significant reductions in all types of pesticides, insecticides, and fungicides (Hyland et al. 2019).

A study of 68,946 adults conducted in France and released in 2018 found that adults who ate more organic food had 25 percent fewer cancers than those that never ate organic. The authors expected to find some reduction in cancer, but they were surprised by the magnitude of the difference (Baudry et al. 2018; Rabin 2018). Other researchers have suggested that the findings be read with caution because they did not test for pesticide exposure and that more research is needed to confirm the results. A previous study also showed a significant reduction in non-Hodgkin lymphoma in those who ate organic food on a regular basis (Mie et al. 2017). There is a strong body of research that shows that at least three classes of pesticides and herbicides that are regularly used in conventional agriculture production are linked with higher rates of cancer (Hemler et al. 2018).

Even if the evidence on nutritional quality of food is still unclear, there is no doubt that eating organic food reduces chemical exposure in the body, and that has significant health impacts. Moreover, those who grow organic food versus those who grow conventional food are exposed to significantly lower levels of pesticides while they work.

Is Organic Farming Better for the Environment?

Although many studies find that organic agriculture offers significant environmental benefits, there are plenty of studies that have found them to be no better than conventional agriculture. The most likely scenario is that there is a range of organic farming styles; some trend toward environmentally sustainable practices, and others do the minimum to adhere to the organic standards (Rigby and Cáceres 2001; Guthman 2004).

Studies on soil ecology confirm that organic farming practices do improve overall soil quality and health (Stinner 2007; Gomiero et al. 2011; Schrama et al. 2018). A study by Mäder et al. (2002) published in *Science* found that organic systems had greater soil biodiversity and were subsequently better at resource utilization (making nutrients accessible to plants) than conventional systems. A common assumption has been that organic systems would be limited by the availability of nitrogen. Several studies have refuted this assumption and shown that over time organic systems were better off than many conventional farms (Stinner 2007; Gomiero et al. 2011).

Organic farms have been found to have a wider range of beneficial insects and arthropods than conventional farms and twice the total number of species (Stinner 2007). These insects and arthropods help keep pests that damage crops in check. Overall, biodiversity is higher on organic farms (Gomiero et al. 2011). Data from the past 30 years consistently shows that organic farming systems increase species richness by 30 percent, and there is an even greater impact on biodiversity if the farm is located in a high-intensity farming region (Tuck et al. 2014). In regions dominated by field crops, this is generally because organic grain and pulse farmers are required to use a multi-year crop rotation, which provides a much broader diversity of habitat than the typical conventional rotation of corn-soybeans (Levins et al. 2017). Including a hay or grazing rotation increases benefits even further.

Water quality improvements shown in organic systems, especially grain crop rotations, are significant. One study showed nitrate pollution from organic farms was as much as 50 percent lower than in conventional corn and soybean farms (Cambardella et al. 2015). The reason for such a large reduction is not merely due to the reduction of synthetic applications of nitrogen, but also due to the remnants of previous crop rotations such as perennial roots from alfalfa or other grasses left in the soil that reduce leaching of nutrients including nitrates.

One significant finding that organic farmers have been working to mitigate is the negative impact that tillage can have on soil. Tillage is a useful mechanism for removing weeds, but it can be damaging to the soil. New research is attempting to find alternate ways to deal with weeds to reduce tillage on organic farms.

Can Organic Farming Help Solve the Climate Crisis?

Human-induced climate change has become one of the largest threats facing agriculture worldwide. According to the National Climate Assessment, average temperatures in the United States have risen from 1.3 °F to 1.9 °F with much of the temperature rise occurring in the past 50 years (Wuebbles et al. 2017). While that may not seem like much of a change, the rate of change in the previous few decades has been significantly higher. The Intergovernmental Panel on Climate Change (IPCC) estimates that human activities that result in greenhouse gas (GHG) emissions have accounted for about 1.0 °F of the increase in average global temperatures. Greenhouse gases are a combination of carbon dioxide, nitrous oxide, and methane.

Since the 1990s, GHG emissions from agriculture have increased 17 percent, due, in part, to increasing use of liquid manure for fertilizing crop fields. The EPA estimates agriculture accounted for 9 percent of GHG emissions in 2017 and is the fifth highest industry source in the United States. Globally, agriculture and deforestation (land cleared for growing crops) account for somewhere between 10 and 30 percent of GHG emissions (Seufert and Ramankutty 2017; Skinner et al. 2019; IPCC 2014). While agriculture is a major contributor to the climate crisis, it is also one of the most vulnerable economic sectors. The climate crisis can put significant stress on all types of agriculture with extreme heat, drought, flooding, or heavy rainfall. These weather changes will increase incidents of weeds and pests, and the extreme weather events will cause water shortages, erosion of soil, changes in growing conditions, and season length (Niggli et al. 2007; Melillo et al. 2014). Many agricultural crops are regulated by climate, meaning the initiation of growth and fruit development depends on specific temperatures for a certain duration. Agriculture will need to adapt to these changing conditions to reduce the economic and food security impacts of the climate crisis (Hatfield et al. 2014). Organic agriculture can play a key role in adaptation and mitigation strategies, but the climate crisis is a complex issue. The climate crisis has the potential to be both a threat to farmers and also a benefit in some cases. In addition, farming can both contribute to and potentially mitigate the climate crisis. In many ways, organic farming can provide solutions to the climate crisis, but not all organic practices are good solutions. Many solutions will need to draw on techniques specifically designed to address the climate crisis regardless of organic status (Niggli et al. 2007; Schonbeck et al. 2018).

Almost half the emissions from agriculture come from various soil management strategies, especially applications of synthetic and organic fertilizers, loss of organic matter through tillage and irrigation, and growing large quantities of nitrogen-fixing crops. Certain management strategies contribute significantly more GHGs than others. Bare soils are subjected to wind and water erosion, and additional tillage breaks down the soil structure even further, allowing carbon to be released into the atmosphere. Overgrazing pastureland is also subject to higher rates of erosion and loss of organic matter. Livestock production practices, especially on a large scale, are the second highest source of emissions in agriculture, including methane produced by ruminants and the management of manure lagoons typical on large conventional farms (EPA 2019). Methane produced by the anaerobic fermentation in ruminant animals, such as cattle, and by flooded rice fields is another major contributor to GHGs. The burning of forestland to create arable cropland creates significant GHG emissions and is a major contributor outside the United States, especially for crops grown in tropical regions.

Although there is limited data to support the argument that organic farming practices have lower emissions, one recent study found a 40.2 percent reduction of N_2O emissions per hectare on organic farming systems. The study suggests that beyond nitrogen input, other soil properties can have an impact on the rate of emissions and that organic farming can be a viable option for GHG mitigation (Skinner et al. 2019). Other studies have found no difference between organic and conventional growing systems. A review article that looked at 46 paired organic and conventional systems found that the organic system only offered a 4 percent reduction in GHG emissions (Clark and Tilman 2017). This assessment was based on a lower nutrient uptake from manure vs conventional fertilizers and did not take other management practices into account. Another review found that out of 121 direct comparisons, 72 showed lower GHG emissions per unit from organic while 49 showed similar or higher emissions (Lee et al. 2015). There is a fair amount of uncertainty and debate about how well and how long carbon can be stored in soil, so many scientists do not consider it a viable option for mitigation (Seufert and Ramankutty 2017). Long-term studies have shown that organic management practices have much higher soil organic matter capable of storing carbon (Niggli et al. 2007; Ghorbani et al. 2010), so the potential is there. The main arguments against organic agriculture methods are that yields are lower (and, therefore, more land must be cultivated in order to produce enough food) and that per unit of crop there are higher energy inputs (Schonbeck et al. 2018). Another major critique is that organic farmers rely more on livestock manure, which many assume is not as readily available to the plant as synthetic nitrogen (Niggli et al. 2007).

Conventional farming practices rely on synthetic nitrogen fertilizers to fix nitrogen in the soil; the manufacture of these products uses massive amounts of non-renewable energy and releases significant CO_2 and N_2O into the atmosphere (Ghorbani et al. 2010). There are a number of ways in which organic farming techniques can sequester carbon beyond reducing the use of synthetic inputs. Any practice that builds soil organic matter is going to increase the carbon sequestration rates, and many required organic agricultural practices already do that. Nonetheless, organic farmers can also use methods that, with the specific goal of carbon sequestration, go beyond what is required by organic certification. Reducing or eliminating tillage is one management technique that can make a big difference, though it is often challenging for organic farmers because they rely on tillage for mechanical weed control rather than chemical pesticides. New strategies are being encouraged including using more cover crops, composting, mulching, and crop rotations. These also increase soil organic matter and minimize the time the soil is left bare. All of these strategies work to build soil organic matter (Altieri and Nicholls 2017). Increasing soil organic matter is important because it not only reduces erosion but also improves soil biota, including mycorrhizal fungi. These fungi are responsible for fixing soil carbon, which is essentially a process of storing carbon in the soil and making it available for plants.

Adaptation to the climate crisis can and should draw on many strategies that are commonly used by small-scale organic farmers, but can be adopted on a larger scale as well. Many organic farming systems are inherently designed to be resilient and adaptable. One key management practice is crop diversification and selecting crops and seeds that are adapted to the local ecosystem. Growing a diverse number of crops can also buffer against crop losses due to varying climate impacts (Altieri and Nicholls 2017; Scialabba and Müller-Lindenaluf 2010). Shifting from large-scale confined animal operations to an integrated livestock production system with intensive rotational grazing can reduce excess nutrient production and improve nutrient use in soils. Creating or rehabilitating functional ecosystems in the margins of farmland or on sub-optimal farmland provides all sorts of benefits including soil sequestration, reduced erosion, and water retention, all of which can buffer the effects of the climate crisis (Altieri and Nicholls 2017).

Does the Organic Industry Have a Role to Play in Social Justice Issues?

When organic farming started to gain momentum in the 1970s, it attracted many people who were interested in social issues. Many of the early organic organizations did include social justice issues in their handbooks and policies. They included guidelines for non-exploitative treatment of farmworkers, livestock, and land. The IFOAM's policies include statements about quality of life, safe working environments, and social and ecological justice throughout the entire food supply chain. When the National Organic Standards were finally written 20 years later, there was no mention of fair treatment for farmers or farmworkers. When members of the organic community asked for it to be included, the USDA declared it was outside their realm of oversight.

As the organic industry grew from small independent organic farmers to large businesses and longer supply chains, food businesses often reaped a larger share of the profits. Many consumers believe that buying organic means they are supporting better working conditions, but it is not necessarily true.

Although that may be a common belief, studies have shown that consumers are not all that concerned about farmworker rights. In a study of consumers' interests in food issues, most were concerned about food safety, nutrition, humane treatment of animals, and environmental impacts before they mentioned human rights issues such as working conditions and wages (Howard 2005).

Farmworkers on conventional farms are regularly exposed to pesticides and herbicides that can cause serious health impacts. Those high levels of chemicals then come home with them on their clothes and skin, which then exposes their families to the same levels (Farmworker Justice 2013). Working in organic fields can bring relief from the direct exposure to chemicals, and many farmworker organizations are advocating for organic production methods.

However, workers on organic farmers are not always treated better in other ways. Many farmworkers on large organic farms have low pay, no benefits, and poor working conditions. Several reports by human rights organizations have found hostile working conditions on large organic farms (Mark 2006). Farmowners say that in order to be profitable they would have to double their prices in order to pay farmworkers a fair wage. In a country where Walmart and Costco are the largest retailers of organic food, it seems unlikely to happen on a large scale. The growth of the organic sector has come with a cost in many cases. For example, the demand for organic tomatoes has contributed to the growth of commercial operations in California, Central America, and South America on land that was once used for subsistence farming. The commercial operations have jeopardized the water supply in many regions and have put small farmers out of business (Rosenthal 2011).

A study conducted in 2006 found that organic farmers surveyed in California were not very supportive of a social aspect being added to the current organic certification (Shreck et al.

2006). Even the farmers who believed it was important to provide good working conditions and fair wages did not want it to be part of the certification process. They stated that the market wouldn't allow for much improvement and that requiring it would create too much financial burden. Although not all organic farmers are willing or able to provide better working conditions, the farmers who did tend to offer better wages and benefits to farmworkers in California were usually organic farmers (Guthman 2004). Since many waged farmworkers are working in the United States illegally, they generally do not have the ability to argue for better working conditions.

While organic doesn't always mean better working conditions, it does correspond to more work opportunities. A study of farms in Washington and Oregon found that organic farmers hired more workers per acre than conventional farms, and they worked more days per year (Finley et al. 2018). Moreover, some farm owners have found ways to make their farms a good place to work. They have created models of farming that build in a diversity of crops and work, which means that available work on the farm is spread out over the course of a year instead of just a few months (Finley et al. 2018). That enables them to retain farmworkers on a long-term basis instead of seasonally. Some have determined that value-added products, such as a fruit farm that also sells jams and pies, creates better jobs and wages. Those farmers who are striving to provide better wages all agree that having a high-quality product that gets a higher payback is the key to earning enough to create a good workplace (Shreck et al. 2006).

There are many in the organic community who recognize that for the organic sector to grow to meet the demand, farmers need to make a viable living and should be able to pay their farmworkers fairly. Despite the large challenges in overcoming this hurdle, some large companies are starting to contract directly with farmers and include fair pricing and other supports in the contracts. A number of organizations decided to collaborate on building an organization, called the Agriculture Justice Program, to focus on social justice issues domestically and bringing the success of fair-trade initiatives abroad to the United States. They have created a Food Justice Certified program that allows businesses to prove they are meeting minimum standards and label their products in this way. In addition, some alternative organic certifications intend to include statements on social justice and eventually include them as part of the certification.

Often, organic businesses that start small manage to keep economic and social justice at their core and provide both farmers and workers with good incomes and working conditions. Despite these efforts, as the businesses grow, they very often get bought out by larger corporations that do not have the same set of values. The companies prioritize profits over fair working conditions and relationships with farmers. Other companies avoid getting bought out, but instead, they turn to investors to grow the business or expand into new markets, and then those investors expect a return on investment regardless of how the values of the company are managed. New business models are emerging that can address these challenges. Several models that have been tried include cooperatives, employee-stock owned, and S-corps, but each of those have their own challenges and are not truly protected from outside investor control or outright purchase. One new model being tested out is a stewardship business model in which a company is held in trust with the only mandate to uphold the mission of the company (Gewin 2019). The Oregon-based Organically Grown Company is now owned by a purpose-driven trust and can never be sold. The switch was made possibly by a social finance lender, another prospect for supporting the social missions of organic food companies. Many organic entrepreneurs who are interested in going this route are reaching retirement age and want to avoid the fate of many other companies that were sold out during succession (Gewin 2019). These business models provide an opportunity for companies to commit to paying their suppliers, who are often farmers, a fair price, which then enables those farmers to pay their workers a fair wage.

While there is considerable concern for fair treatment of farmers and farmworkers, there is another element of social justice that can be considered: access to healthy organic food. With more than 80 percent of households in the United States purchasing organic foods at least some of the time, cost is a common concern for organic consumers (AP 2019; Funk and Kennedy 2016). In 2015, the Consumer Reports conducted a review of organic food prices at most major grocery stores in the United States. They found that organic foods were on average 47 percent more expensive than their conventional counterparts, putting them out of budget for many consumers, though they also found a wide variation in prices across products and retailers. In some cases, they found many items were the same price or even less expensive. The most affordable organic foods were found at Trader Joe's, Wegmans, and Costco (Consumer Reports 2015). As larger retailers increase their organic options, they lower the prices by buying in large quantities. The growth of large food companies and increasing imports is having an impact on the prices of organic foods. That may be good for consumers, but it can come at the expense of farmers earning a living, and those who are living in poverty still can't afford to eat organic food.

To address the issues, many organic food organizations participate in food accessibility programs. They work with farmers markets or individual farms to provide coupons or food dollars to make local organic food more accessible. Some operate urban gardens or farms and train young people to grow food, sometimes in collaboration with a foodbank. Many organic farmers find ways to contribute by donating food, offering gleaning opportunities, or providing work trade options for low-income customers. Furthermore, organic production hotspots or locations with high numbers of organic operations have a positive impact on the local economy. Poverty rates are lower and household income higher in counties with more organic production or organic processing and handling facilities (Marasteanu and Jaenicke 2019).

Barriers to Growth

Many of the current controversies in organic agriculture can be traced back to differences of scale and the underlying difference of opinion on the best way to manage organic farms. As organic went mainstream, many in the sector feared their hard-earned growth would be undermined, and as one small company after another sold out to larger ones, their fears seemed to be coming true. As organic food made its way into large retailers, farm sizes increased and began mimicking many of the same practices common on conventional farms. Consumers and farmers started questioning what could be considered organic. This tension is at the root of most of the major controversies and challenges in the organic sector today.

The second half of this chapter will provide an in-depth review of the major controversies and challenges facing the organic industry in recent years. First, it will take a look at the ways in which the NOP has been a success and the ways in which it has failed or still needs to improve. This will include coverage of the most controversial issues that have been making headline news lately, including cases of certification fraud, allowing hydroponic growing systems to be certified organic, stalling regulatory changes for animal welfare requirements, competing certification programs, and the failure of a quest to create a national checkoff program.

Second, it will discuss the crisis in the organic dairy industry and how that is related to other issues in the organic sector. Third, it will address the lack of access to a secure supply of organic seeds and how that inhibits the growth of the entire organic industry. Fourth, it will examine the bottleneck created by a supply shortage in the organic grain sector. Finally, the chapter closes with a discussion on some ways to support the growth and longevity of organic farming with research and farmer training.

Is the NOP a Failure or Success?

Since the inception of the NOP, there have been those who deeply oppose codifying organic farming practices. Growing discontent among farmers who were originally supportive of the program indicates the extent to which the organic sector has grown. Tensions between corporate organic and small-scale organic are prevalent and create distrust among the many organizations, businesses, and individuals that comprise the organic community.

In recent years, a number of decisions and challenges have made a significant fraction of the organic community lose their confidence in the NOP and NOSB. The tensions that exist between the various factions of the organic community make it especially challenging for the NOSB to truly represent the sector and make decisions that work for most industry participants. Even if the NOSB is able to work through the differences and make good recommendations, they don't have the power to require a rule change. The NOP staff have to be willing to take the NOSB recommendations and act on them. Their willingness to do so shifts and changes with the changes in administration and career staff interests. This leaves the organic community in turmoil as many of those recommendations have the potential to significantly change the trajectory of an entire sector.

Some of the major issues that the NOP is accused of not addressing adequately include the controversial decision to allow soil-less growing in hydroponic and aquaponic systems, their lack of action on animal welfare standards, and lack of enforcement of dairy pasture requirements. To some extent, these issues reveal that the NOP does not have an adequate mechanism for updating regulations at the same rate as innovation is occurring. Most of the recent controversies are around methods of production that were not in existence when the regulations were formed.

Many in the organic community see these issues as further representation of the different scales and philosophies at play in the organic industry. These particular regulation decisions can impact farmers differently depending on the size of the operation. Smaller and even mid-scale organic farmers are easily able to meet some of the controversial requirements and want to see the regulations in place to protect their business. Nevertheless, the few large-scale operations using these methods would have to significantly change their operations in order to meet just the bare minimum standards. Until they are required to do so, the rest of the famers fear they will be undercut and lose business, and in some cases, this is in fact coming true. This is the same tension seen throughout most of the major issues facing the organic sector today.

While most farmers still rely on the USDA organic seal, some are starting to explore other ways of communicating their production methods to consumers and regain some trust in their products. Media reports of some of these issues have led consumers to lose trust in the USDA organic seal. Farmers feel let down by the NOP and the fact that the program no longer reflects their way of farming.

Fraud and Lack of Enforcement

Media reports of fraud and improperly labeled organic products occurred almost since the beginning of the NOP. Enforcement of organic certification violations has been almost non-existent until recently. As of 2006, four years after the first products carried the USDA seal, the USDA had yet to issue any fines or pursue any prosecutions in organic labeling fraud despite receiving 50 related complaints (Lavigne 2006). By 2013, the USDA was receiving about 200 complaints a year, but only 19 fines were issued for misusing the label (Charles 2014). The NOP maintained that it did not have the resources to investigate claims of fraud and that they must rely on the certifiers to do their job.

The risk of fraud is especially high for imported goods that are harder to ensure they meet standards. There are those who claim the current system of certification has built in conflict of interest because the certifying agents are in fact hired and paid for by the food producers and not the USDA (Charles 2014; Whoriskey 2017b). In May of 2017, the *Washington Post* broke a story about fraudulent shipments of organic grain from Turkey, one of the largest exporters of organic food to the United States (Whoriskey 2017b). They discovered 36 million pounds of fraudulently labeled corn and soybeans had been imported and at least 21 million pounds were sold as organic before the discovery was made. Much of the grain coming from Turkey is in fact coming from parts of Eastern Europe and not grown organically. Imports of fraudulent organic grain are causing huge issues for the organic industry especially the organic livestock industry as much of the corn and soybeans imported are destined for organic feed. The United States has seen large increases in the amount of grain imported from places like Turkey, and that has led to lower prices for grain crops produced by organic farmers in the United States.

Pesticide tests are not mandatory for all certified operations, but each certifying agency must test 5 percent of their clients. Organic products from China routinely test to show having high levels of pesticide residues. It is a particularly difficult problem in China because the soil and water there are so contaminated that even if food is grown organically it becomes contaminated from washing water or irrigation. Critics say that because the certifying agencies have control over the testing, they can choose the testing criteria and who and when to test, making it possible to manipulate the results (Whoriskey 2017b). An audit by the USDA's Office of Inspector General indicated that there is no enforcement at the ports of entry to ensure that products labeled as organic are indeed from organic farms. Further, if a shipment is legitimately organic but arrives infested with pests or disease, it may be treated with conventional methods but then still sold as certified organic (Harden 2017).

Fraudulent labeling is a concern for an industry that relies on the authenticity and trust in the certification system and organic label. Consumers need to believe they are getting what they pay for, especially as it often comes with a higher price tag. As many organic farmer and consumer organizations realized that fraud in the organic sector was increasing drastically, they started taking action. A Global Supply Chain Task Force was formed to investigate the issue and determine the best way to detect and prevent fraud. In 2017, a bipartisan bill was introduced as part of the Farm Bill to give the NOP more resources to conduct inspections and implement fraud prevention programs.

The resulting 2017 Farm Bill included provisions that required the NOP to issue rules to address fraudulent organic imports. In response, the USDA proposed three strategies for reducing fraud. One strategy is increasing training and accreditation programs for certifiers; another is improving organic traceability with an organic integrity database and increased reporting measures for imports. The final strategy requires an improvement to the enforcement arm of the NOP. With increased measures to reduce fraud from organic imports, the market for domestic organic grain and other products should increase, but not all fraud is related to imports. As much as 85 percent of complaints are about U.S.-based businesses. The most recent issues have been related to the organic dairy sector not meeting pasture requirements. In addition, the largest domestic fraud scheme, involving $140 million dollars of organic grain over the course of a decade, was detected in 2017. The farmer who organized the scheme was sentenced to 10 years in prison. In the scheme, the farmers grew conventional corn and soybeans and then mixed them with some organic grain and sold it all as organic feed to organic meat producers (Foley 2019).

Despite the new strategies and recent fraud cases going to prosecution, some in the organic industry are not convinced that the USDA is going to keep up with fraud detection, and so they have developed their own programs to prevent and detect fraud. As supply chains are getting more complex, it can be hard for certifying agents and companies to ensure there is no fraudulent ingredients in their products. Organizations that work within the organic sector are urging their members to be vigilant about detecting fraud and tracing the sources of their products themselves rather than relying entirely on inspectors to detect fraud. A new watchdog organization, called Organic-Eye, has been founded to investigate organic fraud. In addition, the Organic Trade Association (OTA) has created a new quality assurance program called Organic Fraud Prevention Solutions, which offers resources and training to enrolled members of the program.

Further on the horizon are new methods of detection that can aid investigators and certifiers in discovering instances of fraud. One such method is a test using isotopes to determine how the plant was grown rather than identifying pesticide residue. The technology for this type of test is still in development, so it is something to watch for in the coming years (Skov Jensen 2019).

The USDA is undertaking more plans to address this issue as well. In August 2020, the USDA published the Strengthening Organic Enforcement proposed rule for comment. The rule proposes fewer exemptions for organic certification, especially those businesses that buy, sell, trade, or negotiate sales of organic products. Further it proposes mandating electronic import certificates and standardizing record-keeping requirements.

Does Hydroponic Production Have a Place in the NOP?

Hydroponic production systems grow plants in a nutrient solution rather than in soil. There are a wide range of styles, from tomatoes grown in large trays filled with water and liquid nitrogen to berry plants grown in individual pots that are filled with alternatives to soil such as peat moss, wood chips, or cocoa bean shells and are fed with liquid nitrogen fertilizers. Some operations are located in climate-controlled greenhouses, while others are outside. Hydroponics are controversial in organics for a number of reasons, including the lack of soil and the prevalence of no transition time being required. It was alleged that large-scale berry producers sprayed fields with herbicides (including glyphosate) and then immediately transitioned to hydroponic production with the berries grown in pots that sat on top of black plastic (Wozniacka 2019). This lack of transition time was allowed under the NOP rules as the growing system was not being sprayed because the pots and tubes to feed the system were organic to begin with (Wozniacka 2019).

Back in 2010, the NOSB recommended banning any nonsoil production systems from organic certification. The USDA declined to implement their recommendation and continued to allow hydroponic operations to be certified and sold as organic. In September 2015, the USDA put together a task force to explore the issue of hydroponics in organic systems. The task force prepared a report in 2016 and came to the conclusion that "management of soil is at the heart of organic production" (McEvoy 2016). Then in November 2017, the NOSB board voted 7 to 8 to not exclude hydroponic systems from certification, although it did vote to ban aeroponic production.

Those who support hydroponics claim that the methods have much higher yields with significantly less land in production, and since they are climate controlled, they provide a more secure food supply. Because they use sensors in small containers, there is significantly less water used per crop. The crops can be grown with extended seasons or even year-round in many cases. Companies say that using this growing method means consumers get food produced without synthetic chemicals and all the other environmental benefits. The lack of a transition period means that they can be scaled up to meet demand quickly. Hydroponic systems are efficient and therefore more profitable (Strom 2016).

Opponents say that it can't be organic without soil; that notion comes from the origins of organic (Charles 2017). The argument against allowing hydroponic systems to be certified comes from the codified definition of organic production, which requires certified organic farms to improve the soil quality. The Organic Foods Production Act states, "An organic plan shall contain provisions designed to foster soil fertility, primarily through the management of the organic content of the soil through proper tillage, crop rotation and manuring." Crops grown hydroponically are no longer part of an ecosystem that intertwines a soil microbiome with the pollinators and the surrounding air and water.

It is not just a philosophical argument either. It also comes down to market share and economic differences. Food grown in soil must adhere to regulations that do not impact those grown hydroponically. Growing in natural conditions means being at the whims of weather and climate, putting traditional farmers at a disadvantage (Flynn 2019). The battle over hydroponics in the organic sector is linked back to the argument that the integrity of the organic label is in jeopardy. Some see this issue as a disagreement over the meaning of organic while others see it as a fight over the market competition. This can be viewed as an issue of scale as most hydroponic operations are very large production systems. The long-standing division between those who believe in the philosophical roots of organic, where soil is the key to successful organic farming, is in direct opposition to those who say hydroponic production can meet demand for organic food while reducing environmental impacts (Strom 2016; Charles 2017).

In January 2019, the Center for Food Safety, along with the endorsement of 13 organic farming organizations and consumer groups, filed a petition with the USDA to request regulations that prohibit hydroponic and aquaponic operations from receiving organic certification (Flynn 2019). As of

June 2019, the NOP clarified the rules for container systems including hydroponics, ensuring that no prohibited substances such as pesticides can be applied to the soil during the three-year transition period even if the crops are grown in containers. As for the issue of requiring the soil to be certified organic, the NOP Deputy Administrator, Jennifer Tucker, insisted that the issue was settled and would not be revisited, meaning that hydroponic growing systems would remain part of the NOP and there would be no transition time for container growing (Karst 2019; Tucker 2019). Canada, Mexico, Japan, and the European Union all ban hydroponics from organic certification (Strom 2016).

Trade organizations such as the OTA have gone through an evolution of support for hydroponics over the years. Historically, they did not support the practice, but they have said with the change of definition from the NOSB they no longer support a ban on hydroponics (Strom 2016). This change in position has created a wedge between the organization and many organic farmers.

On March 3, 2020, the Center for Food Safety and group of organic farmers and organizations filed a lawsuit against the USDA over their decision to allow hydroponic operations to be certified organic. The lawsuit rests on the basis that building healthy soil is a core requirement of the organic standards.

Animal Welfare in Organic Poultry and Egg Production

For consumers worried about animal welfare, many assume choosing organic food is one way to ensure the animals they eat are treated well. Unless consumers are purchasing from a small farmer, it is likely that the organic chicken and eggs they purchased were treated no better than their conventional counterparts because animal welfare has not been regulated under the organic program. *Consumer Reports* conducted a survey in March of 2017 and found that over 80 percent of consumers of organic products believe that organic eggs come from chickens that have access to the outdoors (Whoriskey 2017a).

Organic regulations have historically only specified that animals be raised without antibiotics or growth hormones and that they be fed organic foodstuff. How the animals are raised has been a matter of debate with only vague guidelines provided. There is a requirement that animals be allowed to exhibit their natural behaviors and be given year-round access to the outdoors. Without specific rules clarifying the extent of outdoor access, many egg producers have created covered screened porches and considered them as outdoor access (Curry 2017). This is the heart of the new debate. Most of the organic industry want to see some measure of clarification around animal welfare, especially in the egg industry, and have been advocating for standards that minimize stress on the animals and ensure livestock are allowed to exhibit natural behaviors.

In 2011, the NOSB provided several recommendations to improve consistency and transparency in the organic livestock sector. The USDA's Office of Inspector General also published a report indicating that clarification was needed to ensure consistent rules were applied in the organic poultry sector. Over the course of several years, the USDA received numerous public comments encouraging the USDA to enact rules that require organic poultry farming practices meet consumer expectations. In 2012, the USDA conducted an economic impact analysis and found that giving poultry at least two square feet of living space and more access to the outdoors would considerably increase the cost of egg production. The NOSB reviewed all of the feedback and analysis and provided what they considered a balanced set of recommendations.

Following the advice of the NOSB, the USDA drafted the Organic Livestock and Poultry Practices (OLPP) rule. The rule would have established a number of requirements to improve animal welfare in the poultry sector. It established minimum spacing requirements and quality and type of outdoor space that must be available for organic poultry, clarified how livestock must be handled during transport and slaughter, and specified what physical alterations are allowed. The proposed rule was published for comment in the spring of 2016, with the final proposed rule published in 2017. After several delays and revisions, the USDA withdrew the final rule in the spring of 2018. The USDA claimed that the rule would exceed their statutory authority and their assessment of the benefits and burdens. The USDA received 47,000 comments on their decision to withdraw the proposed rule with the vast majority opposing the withdrawal.

The few who do oppose the rule, often large conventional farming lobby groups, insist that organic welfare standards go beyond the intent and authority of the NOP. They want to ensure that the organic certification standard is only about inputs and not about animal welfare. Some opposed to the part of the rule requiring outdoor access cite concerns about biosecurity. The other major argument against the OLPP comes from those who would have had to invest in new infrastructure to provide more space and outdoor access to animals. The passing of this rule was so important to the majority of the organic industry that the final rule was already a large compromise on the level of standards they desired (Curry 2017).

The decision to withdraw angered many in the organic community, who had been waiting for more than a decade to see this rule passed. Several nonprofit organizations filed a lawsuit against the USDA to challenge their decision to withdraw the final rule. Most recently in 2019, the Ninth District Court of Appeals gave them access to documents that allowed them to continue the lawsuit. The OTA filed their own lawsuit against the USDA over the withdrawal of the OLPP rule. In February of 2019, the U.S. District Court for the District of Columbia ruled that the association does have the legal standing to pursue a court case against the USDA for the decision to withdraw the rule. On October 31, 2019, the OTA continued the lawsuit by requesting a motion for a summary judgment. At the same time, the USDA announced that they were reopening the comment period for the proposed rule for a period of 60 days.

In March 2020, U.S. District Court for the District of Columbia found that the USDA had used a flawed analysis as the basis for withdrawing the proposed rule, and the court ordered the USDA to fix the modeling errors, giving them a 180-day deadline to ensure no further delays would be introduced. On April 23, 2020, the USDA published a request for comment on the Economic Analysis report as part of the court-ordered process to revise the document. The comments will form part of a final report to be published later in the year.

Battle for a National Organic Checkoff Program

Marketing campaigns for a particular commodity are often funded by checkoff programs. These programs require farmers to pay a tax to a national fund that then promotes commodities such as beef, pork, peanuts, and avocados. The USDA manages 22 such programs. In 2002, there was an attempt to create an organic products promotion checkoff program in the Farm Bill. It passed the senate but not the house. The final bill that passed directed the USDA Economic Research Service to prepare a report for Congress on organic farmers' participation in existing checkoff programs, how they benefit organic farmers, and a review of several proposals for managing organic products under current marketing orders and checkoff programs. In addition, the new farm bill included provisions that gave some organic farmers an exemption from conventional checkoff programs.

With the changes in regulation, the OTA saw an opportunity to create the same kind of marketing campaign for organics. In 2015, the OTA petitioned the USDA to create a checkoff program for the organic industry that would form a fund for research and promotion to be managed by the OTA. The proposed program, called the Generic Research and Promotion Order for Organic (GRO Organic), would promote all commodities in the organic sector and was projected to raise approximately $30 million a year. The OTA wanted organics to be recognized as a distinct commodity class based on production methods. They spent three years conducting research on how to best design a checkoff program to meet the needs of the organic sector. They proposed to include both producers and handlers of organic products and to reduce the burden on smaller producers, making the program optional for any entity with less than $250,000 in revenue.

Not everyone agreed with this idea. Many in the industry were worried that it would only serve the needs of large-scale operations and large food corporations, since most farms would be exempt from participation because they were below the $250,000 threshold. They were also worried not just that they wouldn't have a say in how it was operated, but they would have additional paperwork to file with the USDA to receive the exemption. Others were concerned that it wouldn't necessarily help the domestic market grow, but rather help increase organic imports (Douglas 2015).

Problems within other commodity checkoff programs didn't give many people confidence in the plan either. Controversy within other programs seems to mirror the concerns of those opposed to the organic checkoff program. The programs are usually operated by commodity associations that have strong ties with big industry players, and lack of transparency is a common concern.

Opponents to the checkoff organized themselves by creating the No for Organic Checkoff Coalition with over 775 organizations and businesses. They collected a petition with over 1200 signatures. Over the course of two years, the USDA drafted a proposed rule and collected almost 15,000 public comments. Just before the final rule was expected in 2018, the USDA terminated the proposed rule because of the split in support. They were also concerned that any promotion of organic would necessarily be disparaging to conventional agriculture, which is not allowed by the USDA (Siegner 2018). Despite the program not going forward under the NOP, the OTA decided to go ahead with a voluntary checkoff program called Generate Results and Opportunity for Organic or GRO.

Growth of Alternative Certifications

There have always been a number of alternative certifications and labels that organic producers can use to promote their products. Most of these labels and certifications are in addition to the organic certification, and they cover specific issues such as fair-trade, or rainforest friendly, often to denote more information about products grown outside the United States. Some indicate a particular growing system as biodynamic or a single characteristic such as antibiotic-free, GMO free, and cage-free. Some of those attributes are part of the USDA organic certification, but more and more companies are adding them on to their products because consumers are looking for assurances about the origins of products. There are some concerns that these individual labels confuse consumer understanding of the USDA organic label. Regardless, there has been a growing movement to create an alternative or comprehensive add-on to the USDA certified organic label that goes beyond what is included in the USDA version of organic. To make matters even more confusing, there is more than one of these certification strategies underway.

Two such initiatives were launched in early 2018 to address what the creators saw as deficiencies in the USDA organic label and to bring the original meaning and values behind the organic label back to the forefront. One of the major drivers of this new system is the growing concern over the climate crisis and the belief that if organic farmers focus on soil health they can play a role in addressing it. The certifications also address concerns many have had about the inability of the NOP to deal with animal welfare and social justice issues and the recent controversies around hydroponics and animal access to outdoors. Both of those controversies highlight a disconnect between the origins of organic with soil as the main focus and the consumer understanding of organic as something grown without chemicals. This has spurred members of the organic community to take back control of organic in their own way. The two main alternative certifications underway are developed to be used in addition to the USDA organic label, using NOP requirements as a baseline and building from there. In addition, there are other programs in development that will provide even more alternatives.

Incidentally, both initiatives are using the term "regenerative" in the label. The use of this term is intentional, as many in the international organic community have been advocating for making it a key element of organic methods and certification. Regenerative agriculture is an approach that emphasizes ecological systems including carbon sequestration, biodiversity, continuous improvement of soil, and water quality. Many definitions of regenerative agriculture go beyond this to incorporate systems of animal stewardship, farmworker well-being, and economic stability.

One initiative, the Regenerative Organic Certification, launched by the Regenerative Organic Alliance and led by the Rodale Institute, is attempting to build on the concept of continuous improvement. The goal of the new standard is to incorporate all the elements of numerous stand-alone certifications that address sustainability into one certification that emphasizes both ethics and environmental rejuvenation. The creators want farmers to consider the long-term implications of their farming practices and for the industry to consider how their raw ingredients were grown and how those who grew them were treated. The certification requires the USDA organic certification as a base and adds three overarching elements: soil health and land management, animal welfare, and farmer and worker fairness. They created three levels of certification to encourage farmers to continue improving their operations. The system leverages existing certifying agencies that have undergone additional training to become accredited in the new program. A pilot program was launched in 2018 with over 80 participants and continued in 2019 with 22 participants. These pilot programs have provided feedback that is being used to revise the overarching framework before fully launching the program.

The Real Organic Program intends to create more transparency around organic farming practices and to ensure farms are meeting sustainability goals. They aim to increase soil fertility, biological diversity, animal welfare, and community. They are a farmer-created and run initiative funded by donation. The initiative requires farms be certified organic by one of their approved certifying agencies as the first step toward certification. In 2018, they developed seven specific standards that go beyond the USDA requirements, all of them rules that they believe the USDA should be enforcing or requiring. The requirements were revised in 2019 and opened to any interested organic farmers. The certification is designed to have minimal burden on the farmer, and so it is offered for free with a simple online application and then confirmed with an on-farm tour repeated every five years. The certification is aimed at farmers only and not available for any entity beyond the farmgate such as processors or retailers.

Some farmers are pleased that many of their farming practices will finally be recognized by a certification model, and they hope it will help communicate that to consumers. Other farmers are not so sure they want the extra expense and paperwork of additional certifications, especially since a price premium or increased sales are not certain.

While these additional certifications are gaining some popularity among producers, there are many who are concerned that it will only further splinter the organic community and create confusion among consumers. The organic sector is still a relatively small segment of the agriculture and food economy, so there isn't a lot of room for competing certifications. Some would rather see more effort put into improving the systems that already exist; however, those creating the new certifications do not see that as an option and believe the current systems support large agribusinesses that would not benefit from the regenerative approach and, in fact, lobby to water down the USDA organic certification. Another concern is that with multiple approaches being developed it again leaves open the opportunity for large corporations to co-opt the term "regenerative."

There are some other regenerative agriculture initiatives in the works that focus on outcomes rather than specific practices. In those cases, farmers would not be required to have organic certification or even adhere to all organic practices, but instead would need to meet specific criteria such as soil health and carbon sequestration. Many farmers participating in those programs would be allowed to use synthetic pesticides and fertilizers, something that concerns those who developed the concept of organic and regenerative agriculture.

Crisis in the Organic Dairy Sector

Organic dairy sales are the second largest of all organic products in the United States. In just over 20 years, sales of organic dairy and eggs went from $500 million to $6.5 billion in 2018 (McNeil 2019). In 1997, there were just 12,897 certified organic cows (Dimitri and Oberlitzer 2009). By 2016, organic dairy products represented 5 percent of all dairy products sold in the United States. Those products originated from 280,000 certified organic dairy cows on 2500 farms or 3 percent of the total number of dairy cows (Agricultural Marketing Resource Center 2019). While New York, Wisconsin, and Pennsylvania had the highest number of organic dairy farms, the states with the highest number of cows certified are California, Wisconsin, and Texas. Although Wisconsin has the highest number of organic dairy farms, the mega farms in the west outproduce them by 1.5 times with only a handful of farms (Agricultural Marketing Resource Center 2019).

The U.S. dairy sector is one of the most consolidated industries of organic foods. Dean foods acquired several large organic dairy operations (some with over 4000 cows), and by 2004, it owned over two-thirds of the organic sales in grocery stores. The organic milk supply now exceeds demand, and prices for organic dairy farmers have decreased. It is the small organic dairy farmers that are getting hit the hardest as local dairies are replaced by milk shipped in from larger dairy farms (Michel 2018).

For several decades, consumers have been willing to pay a price premium (about 50 percent higher per gallon) for organic milk due to the belief that organic milk is more nutritious and has a better track record on animal welfare. Despite lack of consensus about the nutritional benefits of organic food overall, organic milk has been found to have significantly higher levels of nutrients, including protein and omega-3 fatty acid. More importantly, recent studies have found significantly lower levels of pesticides and antibiotic residues in organic milk (Welsh et al. 2019).

Growth in the dairy sector is slowing significantly, possibly due to consumers shifting toward a more plant-based diet. While the demand for milk is decreasing, more consumers are buying organic cheese, ice cream, and yogurt (Greene and McBride 2015). This shift in demand for organic milk comes at a time when both the conventional and organic dairy farmers are facing a crisis. The increased demand for processed organic dairy products doesn't seem to be enough to counter shifting economics of the organic dairy sector. The current economics are shaped by the history of the dairy industry overall and the development and growth of the organic sector.

A History of Dairy Crises

The dairy sector in the United States has a long history of turbulent pricing wars and crises stretching back to the first "milk war" in New York in 1883 when farmers dumped their milk instead of selling it to protest the low prices they received from distribution companies. As populations became increasingly urban, dairy farmers needed ways to get their milk to consumers while it was still fresh. Milk handlers stepped in to fill that role, buying milk from farmers in the countryside and ferrying it into the cities. The milk handlers controlled the prices, which fluctuated wildly depending on production levels and demand. Beginning in the depression era of the 1930s, the federal government implemented a series of regulations and programs to prop up the dairy industry.

In the 1930s, one such program, the Federal Milk Marketing Orders managed by the USDA, was established to set a minimum price paid to farmers for fluid milk in certain regions of the country. The goal was to stabilize market conditions and ensure a consistent supply of dairy products (Greene 2017).

The result is that prices are still market-driven, but regulated to maintain a minimum. In 1983 and 1990, legislation was passed to allow dairy farmers to create a dairy marketing board and promote milk products to consumers. Despite these efforts, the dairy sector continues to struggle as dairy consumption overall decreases and costs of production increase.

Similar to agriculture in general, dairy production changed significantly in the post World War II era. As rural areas gained electrification, the possibilities for expanding milking capacity greatly increased. Suddenly, many elements of dairy farming could be automated, and refrigerated tanks and trucks made it possible to sell milk in bulk. Everything from feeding to cleaning and milking had options for mechanization (Saucier and Parsons 2014). This significantly reduced the amount of per cow labor and subsequently increased the number of commercial dairy operations. Paired with more affordable grain feed available due to other changes in agriculture, more farmers reduced grazing time and replaced it with confined feeding and housing systems. The result of all of these changes was that total milk production increased by more than 25 percent while the number of dairy cows was reduced by half over a 40-year period (Saucier and Parsons 2014).

Even though production and efficiency increased, the cost of production rose significantly, and many farmers struggled to make ends meet. Often to counter this, they increased the number of cows they milked and sold into high-volume supply chains. Beyond the economic challenges that the dairy sector faced, the new animal management systems of confinement and grain feeding led to more complicated herd health challenges. Many of these challenges were overcome by applying technological solutions such as routine antibiotic use and the introduction of the growth hormone rBGH.

Organic Dairy Standards and Practices

Although a number of dairy farms had been applying organic methods on their farms for decades, they had no options for selling their milk as organic. In the early 1970s, an organization called Organic Growers and Buyers Association, a now-defunct organization in Minnesota, started certifying organic dairy farms. Before the 1980s, there were still very few certified organic dairy farms, and it was even longer before a certified processor was created. Englebert Farms in Nichols, NY, was one of the first farms to be certified in the country in 1984, and it helped set the standards based on their production practices. At the time, there was no organic dairy sector to speak of, so they sold into the conventional market for 10 years until they joined Organic Valley Cooperative. In 1989, a Wisconsin chapter of the Organic Crop Improvement Association (OCIA) created a set of organic standards, and OCIA International adopted them the same year (Saucier and Parsons 2014).

The organic certification standards implemented by the USDA in 2002 stipulate that livestock be fed organic feed and have access to pasture. The amount of pasture time was vague and has been very controversial. The intent of the regulation was to ensure the animals were able to behave as they normally should, and for ruminants, that means eating on pasture as much as possible. The original regulations did not stipulate a minimum time because the grazing season varies according to climate, and a farm in Wisconsin will have a vastly different grazing season from the one in the high plains of Colorado; accordingly, without specific criteria it was difficult to enforce the regulation. Many large dairy farms were providing very little if any pasture time for their herds. Many stakeholders, including traditionally sized dairy farmers (100 cows or so), pushed the USDA to incorporate specific criteria into the regulations. Finally, the "pasture rule," as it is called, was introduced in 2010. It stipulates that all ruminates must receive at least 30 percent of their food intake from pasture for at least 120 days during the grazing season. Despite the change in regulations, it seems some large farms were not always following those rules. Many argue that the large-scale operations cannot possibly meet the minimum standard for pasture access given the number of animals and timing of milking. They were right in believing that the large-scale farms with 3000–5000 cows did not graze their animals on fresh grass. Instead, the animals were kept in large pens outdoors and fed grain and hay. The operators of those farms argued that these conditions were necessary in order to provide a consistent supply of organic milk at affordable prices (Fromartz 2007).

Organic dairy farmers generally milk their cows just twice a day compared to three or four times a day on a conventional farm. This is because organic dairy cows need enough time on pasture. Conventional farmers generally keep their lactating cows indoors and feed them a high-grain diet. The organic dairy cows tend to produce milk for 10 years or more (taking breaks to birth new calves who are often sold to supplement income), while conventional dairy cows burn out much faster and are replaced every few years. Further regulations stipulate that in organic dairy production, hormones and antibiotics are not allowed, although treatment cannot be withheld to maintain organic certification. Once a cow has received antibiotic treatment, it must be removed from the farm and cannot be sold as organic.

Consequently, organic farmers focus primarily on preventative management practices. One study found that organic farmers tended to use crossbred cattle rather than purebred Holsteins, and their herds tended to be older. They were less likely to cull for low production, and as required by NOP rules, they did not dock tails. Many farms used breeding bulls rather than artificial insemination. Likely this is due to the fact that organic farmers tend not to use veterinary care on a regular basis. In fact, many farmers choose organic because they are concerned about chemical poisoning and herd health rather than economics (Sorge et al. 2016). Before converting to organic, many dairy farmers were paying tens of thousands of dollars a year for veterinary treatments, and visits would often be weekly. Once the farmers converted to organic, herd health improved significantly, and the huge costs for treatment were nearly eliminated.

Growing an Organic Dairy Sector

Demand for organic products really began in the 1990s with the advent of rBGH. It became the fastest growing product in the organic sector and quickly became available in mainstream grocery stores. rBGH or recombinant bovine growth hormone is a genetically engineered hormone created by Monsanto under the brand name Posilac to increase milk output in dairy cows. Despite many public safety concerns, the Food and Drug Administration approved rBGH in 1993, and it went on the market in 1994. The synthetic hormone was controversial right from the start, both for the potential impacts to human health and for the impact on the health of the cow. Many countries including the EU, Canada, Japan, and Australia have banned the use of rBGH.

The huge consumer backlash to the use of rBGH led many farmers and processors who did not use rBGH to start noting that on their labels. Monsanto and others in the dairy industry fought back, and thus in many jurisdictions, that information is not allowed on the label. The only way consumers could be sure they were getting dairy products free from rBGH was to buy certified organic products. Thus, the demand for organic dairy grew quickly, especially among parents of young children. Consumers who were not interested in buying organic food suddenly wanted organic milk, and the organic industry hoped this would lead to a greater demand for other organic products.

Before the rBGH-led market growth, organic dairy farmers received very little economic benefit from their organic status. The exception was a small group of individual farmers who chose to create their own small-scale on-farm processing. They bottled milk and made cheese and yogurt. Some farmers banded together to build a processing facility and set up delivery routes, usually selling to small health food stores and food coops. Many of the farmers did not label their products as organic even though they were using organic practices. There just wasn't a strong consumer demand for it in the early years of individual processors during the 1970s and 1980s. There were no large supermarkets that carried independently produced milk, let alone certified organic milk.

Over time, about 50 small regional milk buyers and processors operated throughout the country. Eventually, there were three major national buyer/processors that overtook the industry. Organic Valley, the brand name for the Coulee Regional Cooperative formed in Wisconsin in 1988, was the only organic processor buying large quantities of organic milk for several years. It is farmer-owned and sets member-determined prices. Then came Horizon Organic, founded in Boulder, Colorado, in 1991, which started by processing yogurt with milk bought from Organic Valley and then transitioned into selling fluid milk. In order to meet the demand for fluid milk on the West Coast, they opened a 4000-cow dairy farm in Idaho (Fromartz 2007).

For many years, they operated three large dairy farms with over 2000 cows at each facility, some as many as 4000. Two of those farms have been closed, and they now operate just one in Maryland (Greene and McBride 2015). Horizon was bought by Dean Foods in 2005 and then sold to Danone in 2016. In 2003, one year after the organic regulations went into effect, Aurora Organic was created when Aurora Dairy, which ran CAFO dairy farms across the country, converted one of their farms to organic (Kastel 2018). Over time, Aurora Dairy converted more of their farms to organic and built a processing and packaging facility next to one of their farms. By 2015, they had over 22,000 cows on just five farms (Greene and McBride 2015). Their main focus was on producing private label milk for grocery store chains (Su and Cook 2015).

For many years, the organic sector gave dairy farmers a safe economic shelter from the historically volatile prices. Organic processors would offer longer-term contracts, often 1–2 years, and while prices were market-driven, the demand for organic increased significantly for many years, making it profitable. In addition, Organic Valley implemented a self-imposed quota system that kept prices stable (Su and Cook 2015).

This growth was maintained until 2007 when there was a surplus of organic milk for the first time (Dyck et al. 2009). Much of the growth was driven by large retailers, such as Walmart, Whole Foods, and Trader Joe's, creating their own store brands of organic milk while also selling private label milk brands. The development of store-branded milk, which is usually the same milk that is sold under another brand, complicated the dairy industry. The fast growth in demand supported the development of very large commercial-sized diaries, undercutting prices for family farmers.

Prices paid to organic dairy farmers were profitable for many years, which drew new farmers to transition to organic farming as the conventional prices were quite low and thus led to the resulting surplus (Dyck et al. 2009). This subsequently led to the creation of new manufactured products such as cheese, sour cream, and ice cream; nevertheless, the economic downturn of 2009 had an impact on the dairy industry with sales dropping sharply. Processors and distributors lowered prices, instituted quotas, and asked for voluntary reduction in production. Many farmers who had long-term contracts with verified prices lost those to monthly variable pricing structures. The organic dairy sector was in crisis for the first time since it went mainstream.

Regulatory Changes Address Issues of Scale in Organic Dairy

The growth of the organic dairy industry mirrored the challenges faced in other organic sectors when large corporations entered the market. Small-scale farmers put the health of animals first before profits, and the cooperatives and small processing companies supported their efforts to make traditional-sized dairy farms an economically viable option. This was especially important in an era when the conventional dairy sector was consolidating, leaving few options for farmers to stay in business and make ends meet. With the introduction of mega farms owned by Horizon and Aurora, smaller farmers were justifiably nervous about the future of organic milk production.

In 2005, the Cornucopia Institute, a corporate watchdog and organic farm advocacy group, launched a coordinated effort to tighten the regulations around the requirement for access to pasture. They filed a complaint against Aurora Organics with the NOP claiming the company was not meeting the pasture requirement. The NOP, in return, asked the NOSB to clarify what access to pasture meant. At the next NOSB meeting, nearly 50 farmers testified to support more rigorous pasture standards being put into the regulations. Their testimony was backed up with approximately 8000 letters of support. The NOSB voted to rewrite the pasture regulation to state that cows must graze on pasture during the growing season, and that included lactating dairy cows. A separate guidance document indicated that cows should graze for 120 days per year and only be confined during poor weather conditions. The NOP in turn asked for more clarification in the new regulations. This back and forth negotiation between the NOP and NOSB continued for many years. Eventually, Aurora organic was found in violation of 14 aspects of the organic regulations, but it managed to negotiate a probation rather than decertification (Kastel 2018; Whoriskey 2017a).

In 2015, the USDA proposed to amend the Origin of Livestock requirements for dairy cows by specifying that after transitioning an operation to certified organic production any new animals introduced to the operation would need to be organic from the last third of gestation. After a period of time for comments and investigation, the rule change stalled, and no further action was taken for a number of years. This proposed change was supposed to close a loophole that allowed continual transition of conventional cows into an organic operation. Rather than raising a cow organically from birth, the cows were raised on conventional feed and then transitioned back to organic feed for a year before milking. This practice reduced costs and put farmers who did not use this shortcut at a disadvantage. Some organic certifiers allowed this practice while others didn't. Many in the organic dairy sector were pressuring Congress to ask the USDA to release the final rule and enforce it. On September 30, 2019, the USDA announced they were reopening the comment period on the proposed livestock rule for a 60-day period. They received an additional 600 comments (over 1500 comments were submitted in 2015) with the majority supporting the rule. No further progress has been made.

Current Organic Dairy Crisis

While the dairy sector never quite regained the growth records of the first 20 years, it did recover after the economic downturn in 2008 due to continuing consumer demand and willingness to pay higher prices for organic milk (Walsh 2019). In 2016, demand for organic milk outpaced supply again, but that quickly changed with an expected surplus in 2017. The organic dairy industry once again started showing signs of trouble, with organic milk sales declining and output increasing, resulting in low prices for farmers. By 2018, the pay price of organic milk was down by 25 percent.

Processors responded by implementing production quotas and selling excess milk into the conventional market or by outright canceling contracts (Walsh 2019). The result is economic instability for farmers, and many get forced out of dairy farming altogether. Milk buyers and processors are dropping small local producers in the east and Midwest in favor of the large commercial operations in the western United States where prices are cheaper. Processors offering all or nothing contracts or insisting on expensive upgrades and refusing to pick up small loads means that small farmers bear a disproportionate amount of loss in the market.

For decades, organic dairy farms have avoided the economic cycle of the conventional dairy sector where increased prices lead to surplus of supply, which then leads to a fall in prices and loss of markets, but now the organic sector is falling into that same trap. Organic farmers and experts feel there are a number of issues that are causing this new crisis in an industry used to stability and growth. Many farmers blame the loopholes in the organic regulations that allow large-scale producers to transition conventional cows into an organic herd and a lack of enforcement of the pasture rule. Some blame the decreasing demand for organic milk on the relatively new interest in alternative milks such as soy and oat.

Policy solutions that the USDA could take include enforcing regulations such as the pasture rule and closing loopholes by issuing the final Origin of Livestock rule. Further suggestions include mandates for more certifier audits with mandatory unannounced field inspections (Walsh 2019). Other ideas include clamping down on grain fraud (cheap grain makes it possible for large-scale farms to operate) and introducing a quota system. More regional processing facilities or milk aggregation centers located in areas with many small dairy producers may enable small farmers to stay in business. Some smaller processors, such as Maple Hill Creamery, are sourcing their milk only from cows that are pastured or fed hay, and they are marketing themselves as grass-only milk. Other small processors such as the yogurt company, Stonyfield farms, are entering into long-term cost-plus contracts that guarantee a farm's cost of production is covered along with a reasonable profit margin. In return, the company is assured of a guaranteed supply of milk.

Despite all of these solutions, the organic dairy sector is likely to struggle as long as large CAFOs continue to dominate the milk supply chain. Some innovation and regulatory changes could limit the volatility, but it will need to occur soon, or many more farms could end up going out of business.

Access to Organic Seeds

Seeds are incredibly important for farmers because the right seeds can provide protection against pests and diseases. Seeds also impact the way food looks, tastes, and meets nutritional needs. Plant breeding is the foundation of a strong and stable seed supply. Breeding is how plants can be adapted to regional environmental conditions such as soil type and water availability (Hubbard and Zystro 2016). Plant breeding and seed saving also provide desirable plant varieties that are popular with consumers and chefs such as heirloom varieties. There is a long history in the organic community of seed saving and using open-pollinated varieties. Saving seeds mirrors the ethical aspects of the early organic growers who tried to avoid buying many supplies. Seed saving can reduce the cost of production and reliance on off-farm inputs.

Organic seed sources are also important because the NOP requires organic seed when commercially available. Otherwise, the rules allow conventionally produced, untreated seeds until a suitable organic version is available. Although most organic farmers agree that organic seed is important, some farmers dislike using organic seeds because of the higher costs of organic seed and the lower quality, especially around germination rates (Dillon and Hubbard 2011). There are also concerns that this requirement creates a loophole that discourages progress in creating new organic varieties and investing in an organic seed sector.

There has been considerable disagreement over how organic certifiers should handle the issue of only requiring organic seeds when commercially available because there is no clear method for determining how much effort a farmer needs to put into finding a suitable source. Many certifiers find it a challenge to judge because they are not always aware of what organic varieties are available or suitable for a particular farm. Technically, farmers are not allowed to use price as a reason for requesting an exemption from the organic seed requirement, but many farmers indicate that it is their primary motivation. Most farmers, though, say they require a specific variety that is not available in a certified organic option. There are concerns that producers are taking advantage of the exemption and not making the effort to source organic seeds.

Access to regionally adapted seeds that meet the needs of diverse farming systems does continue to be a challenge for organic farmers. Most seed research and development in the past century has resulted in declining diversity in the seed sector rather than more seed options. Consolidation in the seed sector has further reduced seed availability. Historically, seeds were grown, saved, and shared freely as a common resource. Over the past century, free access to seeds and seed saving has eroded as ownership rules have changed significantly (Howard 2015). In the early years of organic agriculture, there were almost no commercially available seed sources, and while there has been considerable growth, it is still a challenge to find appropriate seeds. The pace of growth in the organic seed sector is nowhere close to meeting the needs of the growing organic industry.

There are some efforts by nonprofits and small-farmer collectives to revive the practice of seed saving and breeding for regional variation to help make more seeds available. The Organic Seed Alliance (OSA) began collecting data on the organic seed sector in 2009 and found that most organic farmers still rely on conventionally produced seeds to some extent. Of those growing field crops, forage crops, and cover crops, 30 percent of respondents were using 100 percent of organic seeds in their operations while only 18 percent of vegetable producers were using all organic seeds. Vegetable farmers were the most likely to have increased the amount of organic seed use over a three-year period (Hubbard and Zystro 2016). One of the most notable trends was that as the size of a vegetable operation increased, the use of organic seed decreased. This is most likely due to the available quantity and quality of seeds for a large-scale operation at a reasonable price, and the same issues can be seen with field crop farmers.

Most plant varieties grown widely today are designed for conventional production rather than adapted to organic systems. Conventional seed production typically involves large amounts of pesticides, and many seeds produced for commercial use are then pre-treated with pesticides. There have been some improvements in the past decade or so with the creation of new regional companies focused on producing high-quality organic seeds and a growing interest from the few regional seed companies that remain in the conventional sector to add an organic component to their business. While research support is still a very small fraction of what is available to the conventional sector, that is growing too, often with partnerships formed between farmers and researchers to develop appropriately adapted seeds. Participatory research programs have now become the norm across much of North America. Farmer participation in formal researcher-led studies is vital to the success of the programs, and farmer-led programs are growing. There are many benefits to farmer-led research programs, especially when conducting variety trials.

Much of this work is being supported by funding from nongovernmental and nonprofit organizations and private foundations, but public funding for organic seed research has gone up in recent years as well. Before 2008, public funding remained below a million dollars, but in 2009, that doubled and continued to rise with the development of the USDA's Organic Research Extension Initiative (Hubbard and Zystro 2016). The vast majority of the funding has been designated for breeding and variety trials. These efforts have led to the creation of new varieties adapted to organic farming systems. There are challenges to this research including finding appropriate locations to grow and test varieties and subsequently produce enough seed and access to starter seeds because of restrictive licenses or bans on seed saving. Many growers are fearful of the tactics that large companies, who have used patents to claim ownership over genetic material, have used even against those who were unintentional in their patent infringement.

There is also a growing interest from organic food companies that are looking for particular products. A number of companies are partnering with organizations, farmers, and researchers to develop new varieties to meet market demand (Hubbard and Zystro 2016). Some of these partnerships can ease the challenge that faces organic seed producers when deciding to scale up production. Seed producers want to be assured there will be a market for the seeds before they invest in their production.

As the demand for organic seed continues to grow, several of the biggest challenges will need to be addressed. One is that seed research is largely undertaken by universities but funded by agrochemical and genetic companies that, until recently, had shown little to no interest in funding organic seed research. As public research becomes privatized, so does control over the intellectual property of seed research. The other major barriers and risks facing the organic seed sector are the consolidated nature of the seed economy and contamination from genetically engineered crops (GE crops or GMOs) and pesticides.

Concentration in the Seed Sector

In farming, seeds can be both an input and a product. Until the mid-1900s, farmers obtained seeds from small seed companies or by saving and sharing seeds. The Supreme Court ruling *Diamond v. Chakrabarty* (447 US 303) in 1980 changed the intellectual property rights rule for living organisms, and large corporations began to hold patents on seeds. This process paved the way for GMO seed development and ownership. Large corporations, often biotechnology or pharmaceutical and chemical companies, bought out smaller seed companies, which consolidated the marketplace and, in the process, phased out thousands of varieties. They also merged with one another or created partnerships to share seeds and genetic information (Howard 2009).

The result is that there are only a few options for farmers to get the varieties they need, less research and development being done regionally, and higher prices. As hybrid seeds (seeds cross-pollinated by humans) became more popular, the need for seed companies grew as hybrid seeds cannot be saved and replanted. As fewer farmers saved and replanted their seeds, the knowledge of how to do it was lost, and buying seeds became the norm. The biggest issues that came from consolidation were the financial resources and political clout that the large companies could put behind legal ownership of plant material through patents, which effectively privatized the seed sector.

The largest consolidation of seed occurred under Monsanto, an agrochemical company that bought out over 50 companies between 1995 and 2008 (Howard 2009). This was a part of their strategy to vertically integrate the entire food chain. Companies such as Monsanto and Syngenta are aggressively pursuing legal restrictions on saving seeds even when there is no genetic patent involved. By 2001, there were only six large companies that controlled 60 percent of the seed market worldwide (Howard 2015). In 2010, both the USDA and the Department of Justice took notice of the anticompetitive nature of the seed sector and held workshops across the country to explore the impacts on the agricultural sector. Despite receiving nearly 20,000 responses, the agencies never took any action to address the issues (Hubbard and Zystro 2016). More recently, the top six companies have merged into just four that still control over 60 percent of the seed market (Howard 2018).

Contamination from GMOs

GMO contamination of organic grains is a huge issue for farmers, especially in the Midwest. GMO crop varieties can contaminate organic fields through unintentional cross-pollination, mixing of seeds before planting, or during post-harvest handling. Organic farmers can have their crops rejected by buyers if they test positive for GMO contamination, and as more crops are contaminated, farmers have a harder time sourcing non-GMO seeds to plant. This pushes the market out of the United States to places where GMO use is much lower. Preventative measures such as broad buffer zones and changing planting times can be costly and are borne by the organic farmers rather than the neighbor who uses GMO crops.

GMO crops are not allowed under organic certification, but ensuring that seeds remain free from contamination is nearly impossible. There is significant evidence that coexistence with GMO crops has failed due to cross-pollination, human error, and lack of regulations to control GMO seed use (Dillon and Hubbard 2011). By 2014, over 90 percent of conventional corn, soybeans, cotton, canola, and sugar beets grown in the United States were GMO varieties (Greene et al. 2016). Sweet corn, squash, alfalfa, and papaya are often grown using GMO varieties.

GMO contamination in canola, soybeans, and corn is widespread in both conventional and organic crops. When organic crops are contaminated, especially seed crops, it can completely eliminate varieties of organic crops. Seed companies are struggling to find and maintain uncontaminated seed stock. Not only contamination but also the pesticides that are associated with the GE crops are an issue with GMOs. Pesticides often drift to nearby farms and contaminate crops, soil, and water and impact the farm's biodiversity.

The contamination is so pervasive that seed companies are hesitant to test for it because there is no recourse. So even seed certified as organic is often contaminated with GMO material, and thus farmers unknowingly grow and sell food contaminated with GMOs. This puts the farmer at risk of losing certification or contracts because their product will test positive for GMO contamination. Unlike pesticides, there is no tolerance level for GMO material derived from drift contamination. This lack of transparency led many in the industry to develop a new voluntary certification in 2007 called non-GMO verified (Thottam 2007) backed by testing for GMO contamination.

This label is now used on thousands of products, often alongside the organic certification label.

Efforts to reduce contamination include using buffer strips or segregating organic fields from conventional ones, keeping organic and conventional products separate through the entire food chain, and regular testing for GMO material. Organic corn producers may stagger planting to avoid pollination during the same time as conventional farmers. These mitigation efforts come at an economic cost for organic producers as they are responsible for keeping their crops GMO free. A study conducted by the USDA ERS showed that in 2014 certified organic farmers lost $6.1 billion in sales due to GMO contamination (Greene et al. 2016).

Contamination from Pesticides

All herbicides have the potential to drift off the intended fields after application, but two herbicides in particular are prone to volatilize for days after application. Dicamba and 2,4-D can move long distances, causing damage to a wide variety of crops including beans, peas, tomatoes, and other fruits and vegetables. Dicamba was developed to control glyphosate-resistant weeds. The pesticide, produced and sold by Monsanto and BASF, has been reportedly damaging much more than just organic farms. In 2017, there were over three million acres of crops damaged by dicamba drift (Dewey 2017). For organic farmers, the loss can be devastating.

Currently, each state is responsible for managing pesticide drift issues, but many states are not tracking instances of drift. There are many anecdotal cases of organic farmers losing a year of crops while pesticide applicators are receiving warnings or minimal fines. State agencies are getting thousands of complaints each year, but there is no way of knowing how widespread the problem is (U.S. EPA 2014). With no federal program to regulate it, there is unlikely to be any significant change. Even when there are state policies in place to deal with the issue, it can be difficult to detect from where the pesticide originated. Many organic farmers only realize there is a problem when their crops start to show signs of pesticide damage. Once a problem has been detected, the farm must retest until the residues are under the acceptable limit. Depending on the severity of the contamination, a farmer may have to re-transition their field or lose certification altogether. Beyond the loss of profits and certification, small farmers may lose their credibility, customer base, and seed supply.

Under the organic standards, organic farmers must take steps to reduce the likelihood of contamination. This can be through physical barriers such as bushes and hedgerows, management practices such as timing of planting, and by communicating with neighboring farmers or putting up signage. The NOP recognizes that pesticides are persistent in the environment and have set up tolerance standards for allowed residues. Samples of a product must not show more than 5 percent of the EPA tolerance level for a given pesticide (Maynard et al. 2019).

There have been some attempts to reduce the problem of pesticide drift beyond regulations. Researchers and staff at Purdue University created a voluntary program called DriftWatch in 2008, which is now run by a nonprofit organization called FieldWatch. DriftWatch is a tool to help farmers and pesticide applicators communicate to try and prevent drift. The tool allows farmers or beekeepers to register their location and the type of crop that should not be sprayed. The tool populates an interactive map that allows applicators to check locations before they apply pesticides in a given region.

Neonicotinoid pesticides (used to control pests such as aphids, spider mites, and stink bugs in orchards and field crops) are also causing issues for honeybees, which are essential for many open-pollinated crops. Beekeepers began reporting record losses of bee colonies in 2006, and now a number of studies have found that the losses are due to exposure to neonicotinoids (Woodcock et al. 2017). Bees are vitally important not just to honey producers, but all farmers who grow pollinating crops, including seed crops. Neonicotinoids can be persistent in the environment, causing long-term effects especially when paired with colonies already stressed from a mite infestation.

More recent studies have shown that neonicotinoids provide very few real benefits to soybean farmers. As of 2015, roughly 40 percent of soybeans planted in the United States had been treated with neonicotinoids, and it is assumed that rate has only increased in recent years; however, the need for the treatment is unnecessary given that most soybeans are grown in a region with few insect pests (Mourtzinis et al. 2019). Furthermore, there doesn't seem to be any relationship between treated seeds and crop yield and subsequently farmer income. When paired with the impact on non-targeted species, there seems to be a strong argument against the heavy use of treated seeds. The regions where soybeans are typically raised also happen to be regions with large migratory honeybee populations, which are necessary for pollinating many other crops.

Possible Solutions

One of the most important ways to protect the U.S. organic seed sector would be better antitrust regulations and enforcement. Another approach would be to eliminate the patent holding of living organisms, although this seems unlikely to occur anytime soon. Shifting plant genetics back into the public realm and out of the private sector would alleviate many of the issues that are challenging organic seed growers and those seeking organic seeds. At the very least, organic advocates would like to see tighter regulations around the development of GMOs, including broadening the definition to include any form of synthetic biology and including an assessment of risks that goes beyond the currently required plant pest risk to include all forms of contamination, related herbicide resistance, and the subsequent impacts on habitat and human health (Beyond Pesticides 2019).

Shifting the burden of responsibility on to those who are growing GMO crops and applying pesticides would help prevent contamination and eliminate the economic loss that organic farmers face when their crops are deemed too contaminated to sell.

Most efforts need to center around supporting independent regional seed supplies by supporting farmers and small seed producers with education, research, and seed-sharing opportunities. Some key recommendations were made by the OSA in their most recent report on how the NOP can support the growth of the organic seed sector. They recommend that certifiers play a stronger role in encouraging farmers to source organic seed, show continuous improvement in using organic seed from year to year, and to encourage larger-scale farmers to contract with seed suppliers for the seeds they need in advance. One way to support this would be to include seed usage into the Organic Systems Plan (part of the certification requirement). Finally, there should be more clarity about how to manage noncompliance and what constitutes noncompliance in the first place. Much of the support to certifiers can come through online tools such as the new Organic Seed Finder database (Hubbard and Zystro 2016).

Overcoming Supply Shortages in the Organic Grain Sector

Growth in the organic sector is hindered in part by the lack of domestic grain and pulse (beans, lentils, peas) crops. Most organic field crop production in the United States is used to feed livestock, including dairy cows and laying hens. Other uses of organic grain include the growing food processing industry such as bread, chips, pasta, and tofu. Organic grain and pulse production have lagged behind other organic commodities, which creates a bottleneck in the processing and livestock sectors. Organic grain producers have been slow to adopt organic practices, and the barriers for new farmers are overwhelming. Organic acreage in the United States is currently just over 5 million acres, but it's still less than 1 percent of the total farmland (Cernansky 2018). Between 2008 and 2016, U.S. organic farmers increased production of corn, soybeans, wheat, oats, and barley by 22 percent. The livestock industry, including dairy, eggs, and meat, grew almost 300 percent in that same period (Short and Molenaar 2018).

In 2016, over half of the organic corn supply and approximately 80 percent of the soybean supply was imported (Levins et al. 2017), a significant increase from previous years. Despite the fact that the United States is generally a net exporter of grains, the organic supply is the opposite. Consumer demand for organic products continues to rise, but because of the lack of domestic grain production, most grains are imported from places such as China, India, and Eastern Europe (Levins et al. 2017). Since a number of organic grain fraud cases have been exposed in recent years, there are many concerns about relying on imports to meet demand. In addition, field crops take up significantly more acreage than any other type of organic production combined, so for those who are concerned with decreasing the negative impacts of conventional agriculture, organic grains and pulses are the most effective way to convert large tracts of land to organic production.

Organic grains and pulses, as currently grown, often have much lower yields than their conventional counterparts, at least in the early years. The cost per unit of production is often more expensive for organic than for conventional crops (Reaves et al. 2019). Long-term trials show that yields are eventually comparable or even better in years with particularly difficult weather stresses, but it takes many years and lots of work to get to that point (White et al. 2019). Eventually, the per unit cost of organic production becomes similar to conventional production. Even before that happens, organic crops are usually profitable because they receive much higher prices (Levins et al. 2017; White et al. 2019).

In order for organic farmers to be profitable in the short term, the price premium needs to be twice the price of conventional crops. Keeping farmers in the market long enough to become profitable even when premiums are not that high is a challenge. This is one reason that so many farmers are concerned about the competition from imported organic grains. As imports increased between 2015 and 2017, prices for organic corn and soybeans dropped significantly (Levins et al. 2017; Short and Moelnaar 2018). These low prices have discouraged new farmers from taking up organic production and reduced the number of farmers interested in transitioning to organic production. In fact, many farmers switched back to conventional production when the prices dropped (Short and Molenaar 2018).

Organic grain production typically requires more crop rotation than conventional production, and it can be difficult for farmers to find buyers for all of the crops in their rotation. Rotations can include anywhere from 3 to 7 different crops all with different buyers and market pressures. This means that farmers must risk growing crops that provide important fertility and weed suppression, but that do not necessarily have profitable markets (White et al. 2019). Since price volatility is an issue in the grain sector, many farmers must choose to sell at a loss or pay to store their crop until the prices stabilize. Either way, it can take many years to recoup the cost and management time necessary for growing organically. This can be especially risky when added to the risk of crops having their organic certification rejected because of GMO or pesticide contamination from neighboring farms.

Recruiting Organic Grain and Pulse Farmers

There are many factors that contribute to the lack of domestic production, but some of the main features include grainbelt farmers' attitudes toward organic production, concerns about profit loss during transition, concerns about weeds and other pests, price of organic seeds and inputs, lack of access to organic production techniques, and other regional variations of barriers (Reaves and Rosenblum 2014). Social pressures and lack of neighborhood support can discourage farmers to take on the risk of trying organic methods.

Grain farmers transition to organic production for a variety of reasons, which have tended to be related to the size of the farm. Small grain farmers generally choose to transition to organic production because they see it as an economical approach to staying small. They already tend to grow more varieties and focus on food grade crops with higher premiums. They also tend to already have a similar mindset about growing techniques and find the transition easier. Finding steady markets for their product and good access to seeds are probably their biggest challenges (Reaves et al. 2019).

Mid-sized farmers who transition to organic production are more likely to transition only a portion of their operation. They try to take advantage of market opportunities, but they are not as likely to stick with organic production, especially if it is not profitable in the short term. Finally, large-scale farmers are the least likely to transition to organic production because it requires the most changes in farming methods and knowledge. They would likely need to entirely revamp their operations, buy new equipment, and learn a completely new way of farming. If the farm is supported or purchased by a large corporation that intends to maintain a stable organic supply, the motivation to stay organic increases (Reaves et al. 2019).

Addressing the variable needs of farmers who transition to organic production is key to encouraging growth in the sector. Most efforts to recruit and train new farmers are focused on small-scale vegetable production as it is the most accessible way to begin. Land access is a considerable barrier for any new entrants to organic grain production, as are the capital costs for new equipment. Recruitment strategies geared toward conventional producers need to address farmer bias toward organics and legitimate concerns about production and marketing challenges.

Supporting farmers through transition and for an additional 3–5 years is an important step in creating long-term organic farmers. It takes considerable trial and error to learn organic farming methods that work on a particular piece of land (Reaves and Rosenblum 2014). This time period is financially risky in that crop yields may be low and certain crop rotations will not have any markets at all. Risk mitigation strategies that support farmers through the transition period might make it more appealing to interested farmers.

Lack of Information and Support

Farmers who are already raising organic crops face challenges with many of the same production issues as conventional farmers, but there is a significant lack of research in ways to deal with them organically. Weeds, soil fertility, and adaptation to variable weather are all major production concerns for organic grain producers. Conventional farmers have access to fertilizer and pesticide consultants and can pick one approach and stick with it. Organic farmers need to use many more tools and a different mindset to grow a high yield crop. Organic farming requires different equipment than is used on conventional farms; in addition, it requires more labor. Access to information on addressing these issues is limited and not often available in the traditional places that farmers would seek out technical information. Many organic farmers leave organic production in the first 3–7 years of production because they do not receive enough mentorship or support (Reaves and Rosenblum 2014).

Weeds may not seem like something that could hold back an entire sector, but for grain farmers, weeds are an ever-present challenge. In order to avoid chemicals and maintain soil fertility, farmers must balance mechanical tillage to reduce weeds with the risk of soil erosion and reduced fertility. Experimenting with different ways to deal with weeds, which vary depending on region, microclimate, and crop type, is an ongoing and time-consuming process.

Research on organic grain production needs to catch up with other production models and be widely available. Access to seeds is an issue for many organic farmers; the lack of quality seeds for production means importing more grains to meet demand.

Possible Solutions

Overcoming the barriers to growing a domestic supply of organic grain will require multiple approaches. Beyond recruiting new farmers and ensuring they have adequate information and support, there are several risk mitigation strategies that can be used to encourage more organic grain production. Historically, conventional farmers were able to mitigate some measure of their risks from year to year by enrolling in crop insurance programs. These programs were structured in such a way that made organic farmers ineligible to fully receive benefits. More recently, the USDA's Risk Management Agency has been working to create a new program specifically designed for organic farmers. In addition, banks are not used to dealing with organic farmers and have difficulty in assessing their value, which means that organic farmers often have trouble gaining access to lines of credit. Farmers often use lines of credit to manage cash flow issues and invest in new equipment.

Flexible market arrangements are a good option for large companies contracting directly with farmers. Supply-chain infrastructure is needed to grow the organic grain sector. Organic grain products must be kept segregated from conventional grains throughout the supply chain, from seeds to final processed products. If access to local grain elevators and processors and transport is limited or non-existent, it will be difficult for farmers who are willing to transition to organic production to find a market for their products. Many large companies that produce processed foods, such as pasta, chips, and crackers, look for individual farmers to contract with instead of relying on the traditional commodity market. This arrangement ensures the company a steady supply of products and the farm is assured of a market. Another way to encourage more gain and pulse production in the United States would be to increase the markets for lower-value crops that are grown primarily to suppress weeds or pests and increase soil fertility.

A new pricing model called cost-plus pricing can relieve some of the pressures the farmers face. This model focuses on the cost of production rather than on the fluctuating market price and can adjust to cover either fixed or variable expenses, plus a margin of profit (Short and Molenarr 2018). This model helps farmers invest in the long-term operation of their farm and encourages farmers to stick with organic production in the early years when yields are lower and the learning curve is steep. It works best when it is built on personal relationships between companies and farmers. Flexible premiums are another way to create a base price or a threshold price that farmers can count on, but still gain the benefit of the market price variations. This can be tied to production criteria such as quality or sustainability measures (Short and Molenarr 2018). Flexible premiums allow the buyer to share the risk with the producer and is best done through direct relationships.

Other companies are encouraging farmers to transition by removing the loss of profit they face during the transition years. Companies are creating transitional labels to support farmers though the first three years (Cernansky 2018). Food companies are also going the route of buying their own farmland to secure a supply of organic grain. Another tactic they are using is following the strategies of large biotech companies and funding research at land-grant universities to improve organic crop production. Beyond that, they are starting to offer technical assistance and planning as part of the purchasing arrangement. This gives farmers the same sort of technical advice they might have traditionally received from conventional input suppliers. Building long-term relationships between growers and those farther along the supply chain can ease the bottlenecks that many food companies are facing (Short and Molenarr 2018).

In 2015, a group of food companies that wanted to expand their organic food production partnered with the OTA to create a U.S. Organic Grain Collaborative, managed by the

Sustainable Food Lab. They hope that by collaborating together, in addition to their individual actions, they can grow the sector by addressing the most pressing barriers.

In some regions, such as the Northeast, farmers are taking advantage of the growing local food economy and have developed a market for heritage grains. These grains are well suited to small-scale operations, organic farming practices, and the regional climate. Farmers are creating unique relationships with brewers, bakers, millers, and others to develop niche products (Blair and Dimitri 2017). This region also has a high number of organic dairies, which further expand market opportunities for grain farmers in the region.

The Midwest offers the most opportunities for growth in the organic grain market simply because that is where the majority of the corn, soybeans, wheat, and barley are currently grown (Reaves et al. 2019). These states also have the most acreage in transition to organic production. Another opportunity is to look at the land in the Conservation Reserve Program (CRP). The CRP pays farmers to set aside cropland for anywhere between 10 and 30 years. When the land becomes eligible to go back into production, it would automatically be certified organic because it would not have been sprayed for many years. Therefore, farmers could begin receiving price premiums right away. The Midwest has a large portion of land under CRP contracts that could potentially increase organic grain production.

Investing in the Future Generation of Organic Farmers

The organic industry faces a number of challenges, and each comes with its own set of solutions that work to address the particular nuances in a region or sector. Some issues require regulatory changes or structural changes at an industry level. There are some common threads of support that can help the entire organic industry, from small farmers to large-scale food companies. The first, and probably most important, is investing in good quality research. The second is making that research available to everyone from educators to the USDA staff and to farmers. Finally, similarly to the agriculture sector as a whole, the next generation of farmers need to be supported through education and training and be given secure access to land and capital.

Organic farming research has been conducted at much lower rates than conventional production, and until more recently, it was largely undertaken outside the traditional research institutes and universities. The main areas of interest in the scientific literature on organic production focus on soil ecology; nutrient management; natural control of pests and diseases; crop quality; and food nutritional quality, composting, weed ecology, and livestock. Other major areas of research include the sustainability of organic production and yield comparisons with conventional farming. Iowa State University established the first organic extension position in 1997 to provide knowledge translation and support to organic farmers and those considering transitioning to organic farming.

As organic production became more acceptable and profitable, corporations started getting involved, mostly by integrating and expanding existing operations. Early research on organic farming was scarce, but by the 1970s, some researchers were showing interest. Lockeretz et al. (1981) published one of the first studies that reported on the economic comparison of organic and nonorganic large-scale farming systems. Production research on U.S. land-grant universities, those dedicated to agriculture research and education, did not begin until much later.

Now, nearly every state has at least some university research dedicated to organic systems, and many universities offer full programs in organic agriculture, but it still lags far behind the amount dedicated to conventional agriculture.

Most organic research is funded by the USDA's Organic Research and Extension Initiative (OREI), which began granting in 2004. Further research is conducted through the USDA Organic Transitions program and the Sustainable Agriculture

Research and Education (SARE) program. An early OREI grant recipient created on online database that consolidates organic research results and communication among researchers and extension agents. A number of organizations are also dedicated to supporting or conducting organic research including the Organic Farming Research Foundation (OFRF) and the Organic Center and many smaller regional organizations.

In addition, many universities and organization are offering education, training, mentorship, and resources to support farmers through transition or farm start-up. Organic farming is knowledge-intensive and requires significant experience to undertake successfully. Many of the organizations reviewed in the profiles chapter of this book are dedicated to providing those necessary opportunities.

References

Agricultural Marketing Resource Center. 2019. "Organic Dairy." *Agriculture Marketing Resource Center*. https://www.agmrc.org/commodities-products/livestock/dairy/organic-dairy.

Altieri, Miguel A., and Clara I. Nicholls. 2017. "The Adaptation and Mitigation Potential of Traditional Agriculture in a Changing Climate." *Climatic Change* 140 (1): 33–45.

Associated Press (AP). 2019. "Good News If You Buy Organic Food—It's Getting Cheaper." *MarketWatch*.

Badgley, Catherine, Jeremy Moghtader, Eileen Quintero, Emily Zakem, M. Jahi Chappell, Katia Avilés-Vázquez, Andrea Samulon, and Ivette Perfecto. 2007. "Organic Agriculture and the Global Food Supply." *Renewable Agriculture and Food Systems* 22 (2): 86–108.

Barański, Marcin, Leonidas Rempelos, Per Ole Iversen, and Carlo Leifert. 2017. "Effects of Organic Food Consumption on Human Health; the Jury Is Still Out!" *Food & Nutrition Research* 61 (1): 1287333.

Baudry, Julia, Karen E. Assmann, Mathilde Touvier, Benjamin Allès, Louise Seconda, Paule Latino-Martel, Khaled Ezzedine, et al. 2018. "Association of Frequency of Organic Food Consumption with Cancer Risk: Findings from the NutriNet-Santé Prospective Cohort Study." *JAMA Internal Medicine* 178 (12): 1597–1606.

Beyond Pesticides. 2019. "Take Action: USDA Must Offer Basic Protection from Genetically Engineered Organisms." *Beyond Pesticides Daily News Blog*, July 16.

Blair, Henry, and Carolyn Dimitri. 2017. "Bridging Crop Diversity and Market Development in the Northeast Grain Renaissance." *Journal of Agriculture, Food Systems, and Community Development* December: 51–60.

Cahill, Stacey, Katija Morley, and Douglas A. Powell. 2010. "Coverage of Organic Agriculture in North American Newspapers: Media: Linking Food Safety, the Environment, Human Health and Organic Agriculture." *British Food Journal* 112 (7): 710–22.

Cambardella, Cynthia, Kathleen Delate, and D. B. Jaynes. 2015. "Water Quality in Organic Systems." *Sustainable Agriculture Research* 4 (3). https://doi.org/10.5539/sar.v4n3p60.

Cernansky, Rachel. 2018. "We Don't Have Enough Organic Farms. Why Not?" *National Geographic*, November 20.

Charles, Dan. 2014. "Can You Trust That Organic Label on Imported Food?" *All Things Considered*. NPR.

Charles, Dan. 2017. "Hydroponic Veggies Are Taking Over Organic, and a Move to Ban Them Fails." *Morning Edition*. NPR.

Clark, Michael, and David Tilman. 2017. "Comparative Analysis of Environmental Impacts of Agricultural Production Systems, Agricultural Input Efficiency, and Food Choice." *Environmental Research Letters* 12 (6): 064016.

Consumer Reports. 2015. "Cost of Organic Food—Consumer Reports." https://www.consumerreports.org/cro/news/2015/03/cost-of-organic-food/index.htm.

Crowder, David W., and John P. Reganold. 2015. "Financial Competitiveness of Organic Agriculture on a Global Scale." *Proceedings of the National Academy of Sciences* 112 (24): 7611–16.

Curry, Lynne. 2017. "Certified 'Organic' Doesn't Say Anything about Animal Welfare." *New Food Economy*, November 9.

de Ponti, Tomek, Bert Rijk, and Martin K. van Ittersum. 2012. "The Crop Yield Gap between Organic and Conventional Agriculture." *Agricultural Systems* 108 (1): 1–9.

Delate, Kathleen, Cynthia Cambardella, Craig Chase, and Robert Turnbull. 2015. "A Review of Long-Term Organic Comparison Trials in the U.S." *Sustainable Agriculture Research* 4 (3): 5.

Delbridge, Timothy A., ed. 2014. *Comparative Profitability of Organic and Conventional Cropping Systems: An Update to Per-Hectare and Whole-Farm Analysis*. Staff Paper.

Dewey, Caitlin. 2017. "This Miracle Weed Killer Was Supposed to Save Farms. Instead, It's Devastating Them." *Washington Post*, August 29, sec. Business.

Dillon, Matthew, and Kristina Hubbard. 2011. "State of Organic Seed." Port Townsend, WA: Organic Seed Alliance.

Dimitri, Carolyn, and Lydia Oberholtzer. 2009. "Marketing U.S. Organic Foods: Recent Trends from Farms to Consumer." Economic information bulletin 58. United States. Dept. of Agriculture. Economic Research Service.

Douglas, Leah. 2015. "Is a National Fund to Promote Organic Produce a Good Idea? Organic Farmers Don't Think So." *Slate Magazine*, May 14.

Dyck, Elizabeth, Katherine Mendenhall, and Northeast Organic Farming Association. 2009. *The Organic Dairy Handbook: A Comprehensive Guide for the Transition and Beyond*. Cobleskill, NY: NOFA-NY.

EPA. 2019. "Inventory of US Greenhouse Gas Emissions and Sinks 1990–2017." EPA430-R-19–001. Washington, DC: US EPA.

Farmworker Justice. 2013. "Exposed and Ignored." https://www.farmworkerjustice.org/sites/default/files/aExposed%20and%20Ignored%20by%20Farmworker%20Justice%20singles%20compressed.pdf.

Finley, Lynn, M. Jahi Chappell, Paul Thiers, and James Roy Moore. 2018. "Does Organic Farming Present Greater Opportunities for Employment and Community Development than Conventional Farming? A Survey-Based Investigation in California and Washington." *Agroecology and Sustainable Food Systems* 42 (5): 552–72.

Flynn, Dan. 2019. "Organic Industry Is Not Giving Hydroponic, Aquaponic Growers a Warm Embrace." *Food Safety News*, February 8.

Foley, Ryan. 2019. "Leader of Largest US Organic Food Fraud Gets 10-Year Term." *US News & World Report.*

Fromartz, Samuel. 2007. *Organic, Inc.: Natural Foods and How They Grew*. Orlando, FL: HMH.

Funk, Cary, and Brian Kennedy. 2016. "The New Food Fights: U.S. Public Divides Over Food Science." Pew Research Center.

Gewin, Virginia. 2019. "Can Mission-Driven Food Companies Scale Up Without Selling Out?" *Civil Eats*, December 19.

Ghorbani, R., A. Koocheki, K. Brandt, S. Wilcockson, and and C. Leifert. 2010. "Organic Agriculture and Food Production: Ecological, Environmental, Food Safety and Nutritional Quality Issues." In *Sociology, Organic Farming, Climate Change and Soil Science*. Vol. 3. Sustainable Agriculture Reviews, edited by E. Lichtfouse. Dordrecht: Springer.

Gomiero, Tiziano, David Pimentel, and Maurizio G. Paoletti. 2011. "Environmental Impact of Different Agricultural Management Practices: Conventional vs. Organic Agriculture." *Critical Reviews in Plant Sciences* 30 (1–2): 95–124.

Greene, Catherine, and William McBride. 2015. "Consumer Demand for Organic Milk Continues to Expand-Can the U.S. Dairy Sector Catch Up?" *Choices*.

Greene, Catherine, Sam Wechsler, Aaron Adalja, and James Hanson. 2016. "Economic Issues in the Coexistence of Organic, Genetically Engineered (GE), and Non-GE Crops." Economic Information Bulletin 149. USDA Economic Research Service.

Greene, Joel L. 2017. "Federal Milk Marketing Orders: An Overview." R45044. Congressional Research Service.

Guthman, Julie. 2004. "The Trouble with 'Organic Lite' in California: A Rejoinder to the 'Conventionalisation' Debate." *Sociologia Ruralis* 44 (3): 301–16.

Hamerschlag, Kari. 2014. "The Assault on Organics: Ignoring Science to Make the Case for Chemical Farming." *Friends of the Earth*, July 6.

Harden, Gil H. 2017. "National Organic Program—International Trade Arrangements and Agreements." USDA Office of Inspector General. https://www.usda.gov/oig/webdocs/01601-0001-21.pdf.

Hatfield, J., G. Takle, R. Grotjahn, P. Holden, R. C. Izaurralde, T. Mäder, E. Marshall, and D. Liverman. 2014. "Ch. 6: Agriculture. Climate Change Impacts in the United States." The Third National Climate Assessment. U.S. Global Change Research Program.

Hemler, Elena C., Jorge E. Chavarro, and Frank B. Hu. 2018. "Organic Foods for Cancer Prevention—Worth the Investment?" *JAMA Internal Medicine* 178 (12): 1606–7.

Howard, Phil. 2005. "What Do People Want to Know about Their Food? Measuring Central Coast Consumers' Interest in Food Systems Issues." *Center for Agroecology & Sustainable Food Systems* 13 (2): 115–25.

Howard, Philip. 2009. "Visualizing Consolidation in the Global Seed Industry: 1996–2008." *Sustainability* 1 (4): 1266–87.

Howard, Philip. 2015. "Intellectual Property and Consolidation in the Seed Industry." *Crop Science* 55 (6): 2489.

Howard, Philip. 2018. "Global Seed Industry Changes since 2013." *Phil Howard*, December 31. https://philhoward.net/2018/12/31/global-seed-industry-changes-since-2013/.

Hubbard, Kristina, and Zystro, Jared. 2016. "State of Organic Seed, 2016." Port Townsed, WA: Organic Seed Alliance.

Hyland, Carly, Asa Bradman, Roy Gerona, Sharyle Patton, Igor Zakharevich, Robert B. Gunier, and Kendra Klein. 2019. "Organic Diet Intervention Significantly Reduces Urinary Pesticide Levels in U.S. Children and Adults." *Environmental Research* 171 (April): 568–75.

IPCC. 2014. "Climate Change 2014: Synthesis Report. Contribution of Working Groups I, II and III to the Fifth Assessment Report of the Intergovernmental Panel on Climate Change." Geneva: IPCC.

Karst, Tom. 2019. "Tucker: Hydroponic Organic Certification Is Settled Issue." *The Packer*. 02.

Kastel, Mark. 2018. "The Industrialization of Organic Dairy: Giant Livestock Factories and Family Farms Share the Same Organic Label." Wisconsin: Cornicopia Institute.

Lavigne, Paula. 2006. "USDA Does Not Always Enforce Organic Label Standards." *Dallas Morning News*, July 25 edition.

Lee, K. S., Y. C. Cho, and S. H. Park. 2015. "Measuring the Environmental Effects of Organic Farming: A Meta-Analysis of Structural Variables in Empirical Research." *Journal of Environmental Management* 162: 263–74.

Levins, Richard, Jeff Howe, Jim Bowyer, Harry Groot, Kathryn Fernholz, and Ed Pepke. 2017. "Organic Grain Production in the Upper Midwest: Status and Prospects." Minneapolis, MN: Dovetail Partners, Inc.

Lipton, Eric. 2015. "Food Industry Enlisted Academics in G.M.O. Lobbying War, Emails Show." *The New York Times*, September 5, sec. U.S.

Lockeretz, W., G. Shearer, and D. H. Kohl. 1981. "Organic Farming in the Corn Belt." *Science* 211 (4482): 540–47.

Lockie, Stewart, Kristen Lyons, Geoffrey Lawrence, and Kerry Mummery. 2002. "Eating 'Green': Motivations behind Organic Food Consumption in Australia." *Sociologia Ruralis* 42 (1): 23–40.

Mäder, Paul, Andreas Fließbach, David Dubois, Lucie Gunst, Padruot Fried, and Urs Niggli. 2002. "Soil Fertility and Biodiversity in Organic Farming." *Science* 296 (5573): 1694–97.

Malkan, Stacy. 2017. "Monsanto Fingerprints Found All Over Attack On Organic Food." *HuffPost*.

Marasteanu, I. Julia, and Edward C. Jaenicke. 2019. "Economic Impact of Organic Agriculture Hotspots in the United States." *Renewable Agriculture and Food Systems* 34 (6): 501–22.

Mark, Jason. 2006. "Workers on Organic Farms Are Treated as Poorly as Their Conventional Counterparts." *Grist*, August 2.

Maynard, Elizabeth, Bryan Overstreet, and Jim Riddle. 2019. "Watch Out for: Pesticide Drift and Organic Production."

McEvoy, Miles V. 2016. "2016 Hydroponic Task Force Report." USDA AMS. https://www.ams.usda.gov/sites/default/files/media/2016%20Hydroponic%20Task%20Force%20Report.PDF.

McNeil, Maggie. 2019. "U.S. Organic Sales Break through $50 Billion Mark in 2018." *Organic Trade Association*. https://ota.com/news/press-releases/20699.

Meemken, Eva-Marie, and Matin Qaim. 2018. "Organic Agriculture, Food Security, and the Environment." *Annual Review of Resource Economics* 10 (1): 39–63.

Melillo, J. M., Terese (T. C.) Richmond, and G. W. Yohe. 2014. "Climate Change Impacts in the United States: The Third National Climate Assessment." U.S. Global Change Research Program.

Michel, Juliette. 2018. "In the US, Small Organic Milk Producers Face Turmoil." *Phys.Org*, March 14.

Mie, Axel, Helle Raun Andersen, Stefan Gunnarsson, Johannes Kahl, Emmanuelle Kesse-Guyot, Ewa Rembiałkowska, Gianluca Quaglio, and Philippe Grandjean. 2017. "Human Health Implications of Organic Food and Organic Agriculture: A Comprehensive Review." *Environmental Health* 16 (1): 111.

Mourtzinis, Spyridon, Christian H. Krupke, Paul D. Esker, Adam Varenhorst, Nicholas J. Arneson, Carl A. Bradley, Adam M. Byrne, et al. 2019. "Neonicotinoid Seed Treatments of Soybean Provide Negligible Benefits to US Farmers." *Scientific Reports* 9 (1): 1–7.

Nelson, Erik, John Fitzgerald, and Nathan Tefft. 2019. "The Distributional Impact of a Green Payment Policy for Organic Fruit." Edited by Jacint Balaguer. *PLoS One* 14 (2): e0211199.

Niggli, Urs, A. Fließbach, H. Schmid, and A. Kasterine. 2007. "Organic Farming and Climate Change." Geneva: International Trade Center UNCTAD/WTO.

Organic Foods Production Act. 1990.

Philpott, Tom. 2012. "5 Ways the Stanford Study Sells Organics Short." *Mother Jones*, September 5.

Ponisio, Lauren C., Leithen K. M'Gonigle, Kevi C. Mace, Jenny Palomino, Perry de Valpine, and Claire Kremen. 2015. "Diversification Practices Reduce Organic to Conventional Yield Gap." *Proceedings of the Royal Society B: Biological Sciences* 282 (1799): 20141396.

Rabin, Roni Caryn. 2018. "Can Eating Organic Food Lower Your Cancer Risk?" *The New York Times*, October 23, sec. Well.

Reaves, Elizabeth, Carol Healy, and Jedediah Beach. 2019. "US Organic Grain-How to Keep It Growing." Sustainable Food Lab.

Reaves, Elizabeth, and Nathaniel Rosenblum. 2014. "Barriers and Opportunities: The Challenge of Organic Grain Production in the Northeast, Midwest and Northern Great Plains." Sustainable Food Lab.

Rigby, D., and D. Cáceres. 2001. "Organic Farming and the Sustainability of Agricultural Systems." *Agricultural Systems* 68 (1): 21–40.

Rosenthal, Elisabeth. 2011. "Questions about Organic Produce and Sustainability." *The New York Times*, December 30, sec. Environment.

Ruskin, Gary. 2017. "Henry Miller's Monsanto Ties." *U.S. Right to Know*, August 30.

Saucier, Olivia R., and Robert L. Parsons. 2014. "Refusing to 'Push the Cows': The Rise of Organic Dairying in the Northeast and Midwest in the 1970s–1980s." *Agricultural History* 88 (2): 237–61.

Schonbeck, Mark, Diana Jerkins, and Lauren Snyder. 2018. "Soil Health and Organic Farming Organic Practices for Climate Mitigation, Adaptation, and Carbon Sequestration." Santa Cruz, CA: Organic Farming Research Foundation.

Schrama, M., J. J. de Haan, M. Kroonen, H. Verstegen, and W. H. Van der Putten. 2018. "Crop Yield Gap and Stability in Organic and Conventional Farming Systems." *Agriculture, Ecosystems & Environment* 256 (March): 123–30.

Scialabba, Nadia El-Hage, and Maria Müller-Lindenlauf. 2010. "Organic Agriculture and Climate Change."

Renewable Agriculture and Food Systems; Cambridge 25 (2): 158–69.

Seufert, Verena, and Navin Ramankutty. 2017. "Many Shades of Gray—The Context-Dependent Performance of Organic Agriculture." *Science Advances* 3 (3): e1602638.

Short, David, and Jan Willem Molenaar. 2018. "Price Management and Investment Mechanisms: Case Studies for the US Organic Grains Sector." Amsterdam: Sustainable Food Lab.

Shreck, Aimee, Christy Getz, and Gail Feenstra. 2006. "Social Sustainability, Farm Labor, and Organic Agriculture: Findings from an Exploratory Analysis." *Agriculture and Human Values* 23 (4): 439–49.

Siegner, Cathy. 2018. "USDA Nixes Organic Checkoff Program." *Food Dive*. https://www.fooddive.com/news/usda-nixes-organic-checkoff-program/523499/.

Skinner, Colin, Andreas Gattinger, Maike Krauss, Hans-Martin Krause, Jochen Mayer, Marcel G. A. van der Heijden, and Paul Mäder. 2019. "The Impact of Long-Term Organic Farming on Soil-Derived Greenhouse Gas Emissions." *Scientific Reports* 9 (1): 1702.

Skov Jensen, Micheal. 2019. "Test for Organic Produce Detects Food Fraud." *Futurity*, August 28.

Sloan, A. Elizabeth. 2002. "The Natural & Organic Foods Marketplace: Mother Nature Moves Mainstream as the Natural and Organic Foods Market Grows Worldwide." *Food Technology Magazine*.

Smith-Spangler, Crystal, Margaret L. Brandeau, Grace E. Hunter, J. Clay Bavinger, Maren Pearson, Paul J. Eschbach, Vandana Sundaram, et al. 2012. "Are Organic Foods Safer or Healthier Than Conventional Alternatives? A Systematic Review." *Annals of Internal Medicine* 157 (5): 348.

Sorge, U. S., R. Moon, L. J. Wolff, L. Michels, S. Schroth, D. F. Kelton, and B. Heins. 2016. "Management Practices on Organic and Conventional Dairy Herds in Minnesota." *Journal of Dairy Science* 99 (4): 3183–92.

Stinner, D. H. 2007. "The Science of Organic Farming." In *Organic Farming: An International History*, 40–72. Oxford: CABI.

Strom, Stephanie. 2016. "What's Organic? A Debate Over Dirt May Boil Down to Turf." *The New York Times*, November 15, sec. Business.

Su, Ye, and Michael L. Cook. 2015. "Price Stability and Economic Sustainability—Achievable Goals? A Case Study of Organic Valley®." *American Journal of Agricultural Economics* 97 (2): 635–51.

Thottam, Jyoti. 2007. "Breaking News, Analysis, Politics, Blogs, News Photos, Video, Tech Reviews." *Time*, March 14.

Tuck, Sean L., Camilla Winqvist, Flávia Mota, Johan Ahnström, Lindsay A. Turnbull, and Janne Bengtsson. 2014. "Land-Use Intensity and the Effects of Organic Farming on Biodiversity: A Hierarchical Meta-Analysis." Edited by Ailsa McKenzie. *Journal of Applied Ecology* 51 (3): 746–55.

Tucker, Jennifer. 2019. "Certification of Organic Crop Container Systems." USDA AMS. https://www.ams.usda.gov/sites/default/files/media/2019-Certifiers-Container-Crops.pdf.

Uematsu, Hiroki, and Ashok K. Mishra. 2012. "Organic Farmers or Conventional Farmers: Where's the Money?" *Ecological Economics* 78 (June): 55–62.

U.S. EPA, OCSPP. 2014. "Introduction to Pesticide Drift." Overviews and Factsheets. *US EPA*. August 1. https://www.epa.gov/reducing-pesticide-drift/introduction-pesticide-drift.

Walsh, Jon. 2019. "Economics of Organic Dairy in New England." *The Maine Organic Farmer & Gardener*, Summer 2019 edition.

Welsh, Jean A, Hayley Braun, Nicole Brown, Caroline Um, Karen Ehret, Janet Figueroa, and Dana Boyd Barr. 2019. "Production-Related Contaminants (Pesticides, Antibiotics and Hormones) in Organic and Conventionally Produced Milk Samples Sold in the USA." *Public Health Nutrition* 22 (16): 2972–80.

White, Kathryn E., Michel A. Cavigelli, Anne E. Conklin, and Christopher Rasmann. 2019. "Economic Performance of Long-Term Organic and Conventional Crop Rotations in the Mid-Atlantic." *Agronomy Journal* 111 (3): 1358.

Whoriskey, Peter. 2017a. "Why Your 'Organic' Milk May Not Be Organic." *Washington Post*, May 1, sec. Business.

Whoriskey, Peter. 2017b. "The Labels Said 'Organic.' But These Massive Imports of Corn and Soybeans Weren't." *Washington Post*, May 12, sec. Business.

Woodcock, B. A., J. M. Bullock, R. F. Shore, M. S. Heard, M. G. Pereira, J. Redhead, L. Ridding, et al. 2017. "Country-Specific Effects of Neonicotinoid Pesticides on Honey Bees and Wild Bees." *Science* 356 (6345): 1393–95.

Wozniacka, Gosia. 2019. "Can Hydroponic Farmers Spray Glyphosate Just Before Becoming Organic?" *Civil Eats*, April 23.

Wuebbles, D. J., D. W. Fahey, K. A. Hibbard, D. J. Dokken, B. C. Stewart, and T. K. Maycock. 2017. "Climate Science Special Report: Fourth National Climate Assessment, Volume I." U.S. Global Change Research Program.

Potatoes
EGGS

Introduction

The essays contained in this chapter highlight the unique perspectives of individuals who are involved with the organic sector to some degree. In the first essay, you will hear from a woman who helped draft the national organic standards about her thoughts on the current backlash facing the program. Next, you will gain some insight on recent developments on a number of critical issues in the organic sector from those who experience it or research it every day. Finally, you will hear directly from a number of farmers about their experience in trying to make a life and a living on an organic farm.

Are the Highest Organic Standards the Best Organic Standards?

Grace Gershuny

In December 1997, the USDA released a proposed rule for implementation of the Organic Foods Production Act (OFPA) of 1990. Long anticipated and also dreaded by the organic community, the draft regulation was soon withdrawn in the wake of a record 280,000 negative public comments. Most of the comments were reactions to the USDA's apparent willingness to allow the use of genetically modified organisms (GMOs) in

Eggs and potatoes for sale at a farmer's market. Seasonal farmer's markets are a popular way for small-scale farmers to sell their products. (Silverblack/Dreamstime.com)

organic production. I was a principal author of that draft and saw the public response to our work as a defeat of the potential of organic farming to radically transform U.S. agriculture.

When I was recruited by the USDA in 1994 to help develop the National Organic Program, GMO crops were in the early stages of commercialization; a key motivation for accepting this job was to ensure that GMOs would never be allowed to carry an organic label. I was elated when the completed proposed regulation, which explicitly barred GMOs from organic status, was approved by top officials at the USDA in June 1997. My elation was short lived. Two months later, the Office of Management and Budget demanded that the USDA delete this prohibition, among other drastic changes.

The outrage engendered by the lack of this explicit prohibition once the proposal was released forced the USDA to change course. The final regulation, published in 1999, prohibits any use of GMOs—now called "excluded methods" —in organic production. This about-face was supported by Monsanto, the chief GMO promoter, who agreed that "consumers deserve a choice"; they also urged the USDA to keep the standards as tough and rigorous as possible to meet consumer expectations for organic products.

Twenty years later, despite the rapid increase in organic sales since implementation of the National Organic Program, only 1 percent of U.S. farmland is under organic management. The vast majority of land remains in agrichemical intensive production, and the major commodity crops—especially corn and soybeans—are overwhelmingly genetically modified varieties.

What Makes Organic Standards Different?

A set of guidelines created in the late 1980s by the newly formed Organic Trade Association (OTA) identified three precepts that set organic standards apart:

1. Organic standards address the process of producing an agricultural product, rather than any measurable quality of the product itself.

2. Organic standards encourage the most environmentally sound farm practices, with flexibility to allow for geographic and site-specific differences, referred to as "agronomic responsibility."
3. Organic standards require producers to demonstrate continual improvement in the quality of their management system, as evidenced by improved soil and water quality, crop quality, biological diversity, and other factors outlined in a farm plan.

The consensus of the organic community supported the first and third of these, but the second provoked a more contentious debate: Should the use of specific farm inputs be allowed or prohibited based on their origin from either natural (e.g., botanical pesticides, mined rock powders) or synthetic (e.g., organophosphate pesticides, anhydrous ammonia) sources, or should the criterion of "agronomic responsibility" be the primary basis for evaluating farm input? Proponents of the "origin of materials" criterion acknowledged that this was neither scientifically valid nor consistent with prevailing norms for organic production. However, they argued that consumers had come to expect that organic food was produced without the use of "synthetic chemicals" and that this expectation should not be violated.

Following a vote by OTA members that narrowly supported "origin of materials," the guidelines were amended to prohibit all synthetic materials and to establish criteria by which some synthetics might be considered acceptable on a case-by-case basis. This mechanism was later adopted in developing the OFPA, which gave responsibility for determining which synthetics should be allowed and which "naturals" should be prohibited to the National Organic Standards Board (NOSB).

The arbitrary distinction between "natural" and "synthetic" inputs set the stage for ongoing fights about the perceived need to uphold the strictest possible standards. Today many passionate organic advocates believe that such rigor is essential for safeguarding organic integrity and consumer trust in the label.

At the same time, many also believe that the USDA's "take-over" of organic standards and the entry of big business into organic markets has brought corruption and weakened organic standards, as evidenced by long lists of synthetic substances approved by the NOSB for use by organic producers.

The assumption is that such loosening of organic standards benefits the USDA's agribusiness clients by allowing big corporate players to enter organic markets with "faux organic" products and to push out the small righteous producers. I refer to this assumption as the "myth of higher standards."

The Myth of Higher Standards

The irony of the organic activist community is that by insisting on higher standards they have given the large professional business organizations an advantage. Bigger operators have the staff and experience to deal with regulatory requirements and the paperwork-heavy demands of organic certification. This has helped create the very situation that organic activists feared the most: intensified bureaucratization of organic certification, increased barriers to access to the organic market by small producers, and near elimination of the possibility that organic production systems might become any more than a small niche in American agriculture.

Much of this confusion arises from ignorance of the difference between standards whose purpose is to regulate markets, such as meat grading standards, and those established to protect public health and safety, such as meat packing plant sanitation standards. Most people—especially environmental activists—are accustomed to thinking of regulations of any kind as serving to protect public health and safety and that industries want the weakest possible standards to minimize their costs. However, the purpose of the USDA's National Organic Program is to regulate the market, a purpose that serves to protect the industry being regulated more than to protect the public.

Uniformity and standardization are ideas that spring from the requirements of industry for inputs and products that are predictable and interchangeable. However, organic systems are characterized by complexity, diversity, local adaptation, and evolution. Food from Florida should not be interchangeable with food from Wisconsin, and the appropriate methods for the organic production of artichokes in California do not resemble those for goat cheese in Maine. By viewing standardization of a process rather than a product, we can recognize that uniform rules do not require uniform practices to comply with the rules.

What Would Truly "Organic" Standards Look Like?

An organic process involves many kinds of products, only one of which will eventually be sold with an organic label. If the ultimate goal of organic methods is the health of the agroecosystem, the quality of the product that is sold is but one indicator of the health of that system.

Performance standards, which establish goals for specific outcomes to be achieved by the operation being regulated, can allow for the kind of innovative problem solving that is a hallmark of organic farmers, and that can be stifled by having to satisfy prescriptive practice standards. In an organic system, measurable improvements in the health of any factor, such as soil quality, crop, or livestock health, can be considered a performance standard.

The insistence on "higher" organic standards works against both creativity and ecological balance. For instance, one important organic precept is that of integrating crops with livestock, both to encourage greater biodiversity and to recycle nutrients within the farm system. However, if a produce grower can only market meat or eggs as organic by following a strict requirement for 100 percent organic livestock feed, she may forego raising her own livestock and buy in manure or compost from conventional livestock or food processing operations. Does this farm adhere to higher standards than one that includes animals who are fed a small portion of local (mostly nonorganic) food waste?

Enforcement is the key to maintaining the integrity of the system, not setting the standards ever higher. Organic certification is legally a license granted by the USDA; before the government can take away a license granted to any citizen, a lengthy process with multiple appeal opportunities must be followed. With the proliferation of various private "beyond organic" labels and activist attacks leveled against the USDA organic program, consumer confusion about the value of the organic label has become widespread. Unlike the federal government, however, such private sector programs are powerless to sanction fraudulent operators with fines and imprisonment.

Consistent standards are necessary for the development of a broader organic market, but this is only one step toward the larger agenda of the organic movement—to transform the way food is produced and distributed in this country. Minimum requirements for making such a shift should be attainable by the majority of farmers and manufacturers who, by definition, will just be "pretty good." One goal of our education system, for example, is to make sure that every child is able to complete high school and to achieve basic competencies. As one organic leader puts it, "Do we want organic farmers to only be those who get the As?"

Grace Gershuny is widely known as an author, educator, and organic consultant. In the 1990s, she served on the staff of the USDA's National Organic Program, where she helped write the regulations. She learned much of what she knows through her long-time involvement with the grassroots organic movement, where she organized conferences and educational events and developed an early organic certification program for the Northeast Organic Farming Association (NOFA).

Different Production Practices for Different Scales of Poultry Operations

Mark Keating

Poultry, including broiler chickens, layer hens, and turkeys, has a complicated place in the organic world. The researchers and practitioners who conceptualized organic farming in the early 20th century embraced nature as the ideal model for an enduring and productive agriculture. Drawing on scientific data and empirical evidence, they contrasted the efficiency and balance inherent in natural systems with the disruption they associated with mechanized farming. For these pioneers, organic agriculture meant using solar energy to recycle natural plant and animal materials through a closed loop (the farm) that yielded abundance and eliminated waste.

What makes modeling nature more challenging when setting organic standards for poultry production? While it is easy to overlook, chickens and turkeys are omnivorous predators with beaks and claws used in the wild to catch and consume insect and animal prey. That's quite different than plants drawing nutrients from fertile, microbiologically active soil or ruminants feeding on a smorgasbord of grasses and grains. Under stressful conditions, chickens and turkeys will use those beaks and claws against each other to impose a pecking order.

Adapting this unique biology and behavior to organic standards for commercial scale production leads to raising poultry in ways that differ greatly from their life in the wild.

The most fundamental modification seen in organic poultry production is transforming chickens and turkeys from omnivores into vegetarians. Part of this transformation involves adding a synthetic version of the amino acid methionine to their feed ration. Natural methionine is abundant in the insects and small animals that poultry hunt in the wild and also at lower levels in the grass seeds they consume. Methionine is essential for healthy growth and feathering in poultry, and deficiency produces cascading effects of poor feed conversion and lethargic behavior. Confined chickens and turkeys experiencing protracted methionine deficiency will attack their weaker companions and consume their flesh and blood to obtain it.

Roughly 50 years ago, commercial agriculture began replacing the natural methionine sources commonly used in poultry feed, such as bone and blood meal, with newly available synthetic compounds. Synthetic methionine was cheaper to produce and did not risk turning rancid. Feed rations continued to incorporate plant-derived methionine from ingredients such as corn, soybean, and wheat by-products, which provided other essential nutrients as well. However, synthetic methionine almost completely replaced the portion previously supplied by animal protein sources. So widely accepted has this changeover become that "100 percent vegetarian feed" marketing claims have become standard on poultry packaging despite chickens and turkeys being anything but vegetarian.

Market demand for organic poultry products did not grow appreciably until after the USDA certification program launched in February 2002. The organic world was so unfamiliar with poultry at the time that there was as of yet no legal allowance for synthetic methionine. The federal approval process had begun belatedly and would not be finalized for another 18 months. In the absence of a viable alternative during the interval, certified poultry farmers continued to feed synthetic methionine. Other provisions in the standards prohibited feeding poultry or mammalian slaughter by-products, so the traditional blood and bone meal were off-limits. Feeding them would also undermine the ubiquitous "100 percent vegetarian feed" marketing claims. Providing sufficient methionine exclusively from grain and seed by-products results in overfeeding the other nutrients they contain.

Growing public awareness of the use of synthetic methionine prompted numerous organic advocacy organizations to campaign for its prohibition. Believing in general that organic standards should allow as few synthetic materials as possible, these organizations view synthetic methionine as a crutch that enables the certification of industrial-style poultry production. They argue that a combination of traditional and novel natural methionine sources can indeed support healthy poultry development. They have advanced alternatives including fish and insect meal, potato starch, and corn varieties bred for higher methionine content in hopes of eliminating the allowance for synthetic sources. These arguments have yet to prevail, and organic standards continue to allow converting omnivorous chickens and turkeys into vegetarians by adding a synthetic amino acid to their feed.

Organic poultry also experience significant modifications to their natural living conditions, and these modifications are magnified on larger farms. Smaller farms raising several hundred to several thousand chickens or turkeys often provide unfettered access to pasture during daylight hours. Poultry on these farms are typically confined overnight, when in nature chickens and turkeys turn from predators to prey and roost in trees for protection. These operations require significant pasture acreage, as poultry must be moved regularly to new pasture to enable soil and plants to recover and mitigate disease risk. Even on these smaller operations, farmers supplement the feed ration with synthetic methionine as the natural sources available in pasture are too limited and inconsistently available to sustain the flock.

Scaling up for organic farms raising 10,000 turkeys and 20,000–25,000 broiler and layer chickens (even larger operations exist) requires far more drastic modifications to their living conditions. Organic standards permit interior stocking levels as low as 1.5 sq. ft. per chicken with comparable densities for turkeys based on their weight. Flocks of this size in such close quarters exacerbate stress, and farmers rely on beak trimming to reduce damage from pecking. Performed before poultry are 10 days old, this practice uses a hot knife or infrared light to remove roughly one-third of the upper and lower beaks and blunt the remaining tip.

Living conditions outside the barn for organic chickens and turkeys also become more problematic with increased scale of production. Organic poultry are required to receive outdoor access regardless of flock size, but the largest operations provide only a screened-in porch that prevents contact with the soil. Large farms that do provide organic poultry with contact to the soil typically see that only a small percentage of the chickens or turkeys in the barn actually venture outdoors. These farms are required to provide outdoor access, not to make poultry go outdoors. There are also multiple conditions including inclement weather and the risk of disease transmission for which outdoor access for the entire flock can be denied.

No doubt the organic pioneers would be surprised by the ways in which poultry's natural behavior and biology have been modified for today's standards. Passionately committed to nature as guide, hopefully they wouldn't judge us too harshly for the allowances made to expand organic poultry production to larger scales. What's most important is that consumers understand the fundamentals of organic poultry production and the fact that management practices vary greatly between certified farms. Certification can provide baseline differentiation between organic and conventional poultry products, but consumers looking for more details, particularly with regard to outdoor access, will need to dig deeper.

Mark Keating started as an organic farmworker in 1988 and has since worked as a cooperative extension agent, university lecturer, policy advocate, journalist, and an organic inspector. He served as the lead organic crop and livestock specialist with the USDA National Organic Program between 1999 and 2002 during the implementation of its organic certification program. He currently works as a consultant on organic, local, and sustainable food projects.

The Politics and Policy of GMO Labeling: A U.S. Perspective

Samantha L. Mosier

It is only since the 1990s that consumers have needed to question if the food they eat is genetically modified or, rather, genetically engineered. The first food product produced as a genetically modified organism (GMO) was the FLAVR SAVR tomato in 1992 followed by the production of Bt corn later in the decade (Bruening and Lyons 2000; EPA 2002). Since that time, interest in regulating GMOs in the food supply has persisted with an increased interest in informing consumers. From a consumer perspective, it is seen as desirable to label food products that contain genetically modified organisms for a range of reasons related to environmental impacts, food safety concerns, exacerbated economic inequality, and general morality implications for "playing God" (Scott et al. 2016; WHO 2014). There is no singular reason behind the call to label GMOs, but the label would provide an informed choice by correcting for a market information asymmetry. Specifically, consumers do not have access to all pertinent and relevant information regarding a food product. A food labeled as containing GM ingredients would provide consumers with additional information about the product prior to purchasing and consuming.

In the past decade, polls have shown that between 70 and 95 percent of Americans prefer GMO foods labeled, but this does not necessarily translate into a decline of purchases or consumption (Annenberg Public Policy Center 2016; Center for Food Safety 2015). Of the possible methods for labeling, Americans express preferences for a clear on-package label through either a text-based indication (e.g. the product contains genetically engineered ingredients) or a symbol. Obtaining the information via a QR code or by calling a 1-800 number is not acceptable. The United States is one of the last countries with a developed economy to adopt a GMO-labeling policy. It is currently one of 66 countries globally to mandate GMO labeling in some form with three countries having outright bans on GMO production and GMO foods (Center for Food Safety 2018).

The movement to label GMOs in the United States was initiated at the state level prior to federal interest in the issue. Indeed, a total of 149 GMO-labeling bills were proposed in 36 states from 2011 to 2016, which marks a period of increased issue salience. Only four states successfully adopted labeling laws. Connecticut became the first state to adopt a GMO-labeling policy in 2013, followed by Alaska, Maine, and Vermont in the years that followed. However, Vermont's GMO-labeling policy was the only compulsory law, which required all food products to be labeled without contingencies of other states adopting similar policies. Vermont's law was able to withstand litigation-based challenges from opponents who claimed the law violated the U.S. Constitution's First Amendment and Commerce Clause (Dillard 2015).

The U.S. Congress finally adopted a national requirement in 2016 (S.764) preempting Vermont's law, but it was generally seen as an unhappy compromise with no one particularly happy with the outcome. The disappointment of the national GMO-labeling law is reflective of the general division regarding the perceived risks and overall safety of GMOs. Democrats argued for mandatory on-package text or symbol labels, which are consistent with expressed consumer preferences. Republicans in Congress, and some Democrats from large agricultural states, were largely against labeling of GM foods. However, if labeling were to occur, they preferred it to be voluntary only and generally supported that consumers obtain the information from a QR code. They claimed consumers are generally misinformed about GMO safety and also that there were high economic costs and implications for producers, processors, manufacturers, and retailers.

Indeed, researchers have found consumers to be incredibly uninformed. Indeed, 33 percent of survey respondents in one study believed a tomato labeled as non-GM meant it did not contain DNA (McFadden and Lusk 2016). Given such evidence, opponents to labeling wonder if consumers would really be able to benefit from clear on-package labeling of GM foods. Instead, other third-party certifications can provide information regarding GM ingredients to consumers. Foods certified as the USDA Organic or labeled as "Non-GMO Project Certified" already are GMO free.

In the end, the U.S. GMO-labeling law was a mandatory labeling law that permitted the USDA to determine how best to require on-package labeling. As a result, the 2018 final rule

(7 CFR Part 66—National Bioengineered Food Disclosure Standard) received a lukewarm response. Proponents for labeling believe the current law permits too many loopholes for companies to hide the GMO content of food, and they even suggest the on-package symbols developed by the USDA are pro biotechnology instead of representing a neutral position. Indeed, not all products containing GMOs would be required to have a label, and the label could be a QR code or 1–800 number. This translates into a number of products that could contain GM ingredients but would not have to carry the label.

Considering the rule was issued under a pro-industry and Republican administration, it is likely the political debate regarding GMO labeling is far from over. It is clear that consumer preferences were not adequately represented in the 2018 final rule. While it is too early to tell what will occur when the USDA final rule goes into full effect in 2020, there is the possibility of litigation to challenge the law or, depending on the results of the election year, a tide of new lawmakers interested in changing the existing policy.

References

Annenberg Public Policy Center. 2016. "Genetically Modified Food Labeling Survey, May 17–May 21, 2016." The University of Pennsylvania.

Bruening, G., and J. M. Lyons. 2000. "The Case of the FLAVR SAVR Tomato." *California Agriculture* 54 (4): 6–7.

Center for Food Safety. 2015. "U.S. Polls on GE Food Labeling." https://www.centerforfoodsafety.org/issues/976/ge-food-labeling/us-polls-on-ge-food-labeling.

Center for Food Safety. 2018. "International Labeling Laws." https://www.centerforfoodsafety.org/issues/976/ge-food-labeling/international-labeling-laws.

Dillard, John G. 2015. "Recapping round 1 of the Vermont GMO-labeling lawsuit." OFW Law Blog Review. https://

www.ofwlaw.com/2015/05/04/recapping-round-1-of-the-vermont-gmo-labeling-lawsuit/.

EPA. 2002. "EPA's Regulation of *Bacillus thuringiensis (Bt)* Crops." Report 735-F-02-013.

McFadden, B. R., and J. L. Lusk. 2016 "What Consumers Don't Know about Genetically Modified Food, and How That Affects Beliefs." *Faseb* 30: 3091–96.

Scott, S., Y. Inbar, and P. Rozin. 2016. "Evidence for Absolute Moral Opposition to Genetically Modified Food in the United States." *Perspectives on Psychological Science* 11 (3): 315–24.

WHO. 2014. "Frequently Asked Questions on Genetically Modified Foods." Areas of Work, Food Safety.

Samantha L. Mosier, PhD, is an assistant professor in the Department of Political Science at East Carolina University. Her research focuses on food and agriculture policy and subnational sustainability initiatives.

Plant and Animal Breeding: A Keystone for Success in Organic Agriculture

Travis Parker

Living organisms display an incredible diversity in their environmental adaptations. Many individual species of crops and livestock, for example, are produced in regions ranging from the humid tropics to chilly temperate and even subarctic climates. But the crop varieties and animal breeds raised in each of these environments vary dramatically from region to region. Like tropical and temperate environments differ, organic and conventional agriculture each provide unique challenges and opportunities for plants and animals. Organic crops are grown without the use of synthetic fertilizers and pesticides. Organically raised animals cannot be treated with non-therapeutic antibiotics or hormones. They also tend to be raised on fodder with lower caloric density and in environmentally variable conditions (FAO 2015). Breeding experiments have shown that the best varieties in conventional agriculture are not always superior when raised organically (Van Bueren et al. 2011; Murphy et al. 2007; Nauta et al. 2006; Wallenbeck et al. 2009). These experiments have shown that crop varieties selected in organic conditions have yields between 5 and 31 percent higher than those bred in conventional systems, when both are raised organically (Murphy et al. 2007). By comparison, total organic yields are approximately 20 percent lower than conventional yields (De Ponti et al. 2012). This implies that organic agriculture could match the yields achieved by conventional agriculture simply by developing varieties adapted to its conditions. But is this happening?

Currently, approximately 95 percent of organically grown crops were bred in conventionally managed environments (Van Bueren et al. 2011). As of 2019, the four largest seed companies are Bayer, ChemChina/Syngenta, Corteva (Dow/DuPont), and BASF (MacDonald 2019). These agrochemical firms have a strong vested interest in maximizing sales of synthetic agricultural inputs. Many smaller seed companies also focus on conventional markets. This has resulted in a lack of resources allocated for breeding in the organic sector.

Conventionally managed environments are doused heavily with synthetic fertilizers, leading to soils with high levels of nitrogen, phosphorous, potassium, and other nutrients. In these conditions, there is little incentive for plants to use nutrients efficiently, and this important trait has deteriorated in some modern crop varieties (van Bueren et al. 2011). Crops must also compete against numerous animal and microbial pests. These problematic organisms are plentiful in nature, and many species have genetic variants that convey resistance to them. But in conventional systems, these are routinely controlled with synthetic pesticides. Some modern varieties of wheat, for example, have increased susceptibility to *Septoria* and *Fusarium* fungal pathogens compared to older varieties

(Simon et al. 2004; Klahr et al. 2007), as selection has prioritized other targets.

Weeds are also a major burden for organic farmers, who cannot control them with synthetic herbicides. Manual weed control in processing tomatoes, for example, requires five times more labor on organic farms (30 hours/acre) than on conventional farms (6 hours/acre) (Klonsky 2011). Because organic weed control is often expensive, crops that compete against weeds effectively through rapid early-season growth are in strong demand by organic farmers. Despite this, the trend in many breeding programs has focused on a reduction in plant size, which can be problematic in environments with high levels of soil nitrogen (van Bueren et al. 2011), and many modern crop varieties show reduced early season vigor and weed competitiveness than their pre-herbicide counterparts. To maintain economic sustainability, organic agriculture will need to move away from crop varieties that were bred to depend on synthetic pesticides and fertilizers, and it will require varieties that have been selected to succeed in organic conditions.

Organic farmers also have different livestock needs than their conventional counterparts. In general, organic farmers require robust varieties that demonstrate strong growth and production traits, even in relatively adverse environments (FAO 2015). This contrasts sharply with conventional agriculture, where animals are typically provided higher energy-density feeds, and they are bred for performance under optimal growth conditions of large operations. Studies on dairy cattle and pigs have confirmed that genetically different animals are optimally suited to conventional and organic environments (Nauta et al. 2006; Wallenbeck et al. 2009), indicating that selection in organic systems will be useful for improving organic animal production.

Fortunately, there is growing interest in improving varieties for organic farmers. In the last decade, public and private organizations have provided increasing support for breeding in the organic sector. This has led to new variety development across many species and environments. The UC Davis organic dry bean breeding program, for example, which has been funded by the Clif Bar Family Foundation, Lundberg Family Farms, the Organic Agriculture Research and Extension Initiative, and Western Sustainable Agriculture Research and Education grants, has developed several new varieties of dry beans for the organic sector. Selecting for disease resistance and maintaining desirable seed colors resulted in 79 percent higher yields. These yield improvements can then be passed on to consumers as lower food prices. Beyond this, additional varieties were bred by selecting for high yields in organic conditions, as well as the previously mentioned traits. On average, these varieties had yield increases of 217 percent, a three-fold improvement over the comparable parent. The new types are planned for commercial release in 2020. The results of this breeding program, and many others like it, indicate that there is enormous untapped potential for organic agriculture, which can be realized through continued plant breeding.

The current lower productivity of organic agriculture keeps prices for organic goods high, making it harder for the general public to access them. Breeding experiments have shown that simply selecting new varieties in organic conditions can improve yields to levels comparable to those of conventional agriculture. More work needs to be done in both the public and private sectors to realize these possibilities. As more varieties are released that serve organic farmers, the costs of organic goods will become more similar to conventional agriculture, improving their accessibility in new markets. This will then serve to increase the economic incentive for further organic breeding. In turn, small investments in plant and animal improvement will be crucial for maximizing the productivity and profitability of organic agriculture in years to come.

References

De Ponti, Tomek, Bert Rijk, and Martin K. Van Ittersum. 2012. "The Crop Yield Gap between Organic and Conventional Agriculture." *Agricultural Systems* 108: 1–9.

Food and Agriculture Organization of the United Nations (FAO). 2015. "Animal Husbandry in Organic Agriculture." *Technologies and Practices for Small Agricultural Producers* 2015: 8378.

Klahr, Anja, Gerhard Zimmermann, Gerhard Wenzel, and Volker Mohler. 2007. "Effects of Environment, Disease Progress, Plant Height and Heading Date on the Detection of QTLs for Resistance to Fusarium Head Blight in an European Winter Wheat Cross." *Euphytica* 154 (1–2): 17–28.

Klonsky, K. 2011. "Comparison of Production Costs and Resource Use for Organic and Conventional Production Systems." *American Journal of Agricultural Economics* 94 (2): 314–21.

MacDonald, James. 2019. "Mergers in Seeds and Agricultural Chemicals: What Happened?" *Amber Waves,* USDA-ERS, 1–7.

Murphy, Kevin M., Kimberly G. Campbell, Steven R. Lyon, and Stephen S. Jones. 2007. "Evidence of Varietal Adaptation to Organic Farming Systems." *Field Crops Research* 102 (3): 172–77.

Nauta, W. J., E. W. Brascamp, R. F. Veerkamp, and H. Bovenhuis. 2006. "Genotype Environment Interaction between Organic and Conventional Dairy Production." https://www.louisbolk.org/downloads/1965.pdf.

Simon, M. R., A. J. Worland, and P. C. Struik. 2004. "Influence of Plant Height and Heading Date on the Expression of the Resistance to Septoria Tritici Blotch in Near Isogenic Lines of Wheat." *Crop Science* 44 (6): 2078–85.

van Bueren, E. T. Lammerts, Stephen S. Jones, L. Tamm, Kevin M. Murphy, James R. Myers, C. Leifert, and M. M. Messmer. 2011. "The Need to Breed Crop Varieties Suitable for Organic Farming, Using Wheat, Tomato and

Broccoli as Examples: A Review." *NJAS-Wageningen Journal of Life Sciences* 58 (3–4): 193–205.

Wallenbeck, A., L. Rydhmer, and N. Lundeheim. 2009. "GxE Interactions for Growth and Carcass Leanness: Re-ranking of Boars in Organic and Conventional Pig Production." *Livestock Science* 123 (2–3): 154–60.

Travis Parker has a PhD in plant biology from the University of California, Davis. He has a BS in Biological Science from California Polytechnic State University, San Luis Obispo. He has been the team lead of the UC Davis Organic Common Bean Breeding Program since 2014. Travis is currently conducting research on genomics.

Early Organic Systems Research
Gigi Berardi

My definition of "organic," the one that I use in *FoodWISE* (2020, 28), is that organic farming uses natural processes for fertilizing and maintaining the structure of soil. It also avoids the use of unnatural (manufactured) substances such as synthetic pesticides.

When I was a young graduate student at Cornell University (in the College of Agriculture & Life Sciences) in the early 1970s, for me to say that I was working on organic farming invited dismay, if not ridicule. The standard rejoinder would be "'Organic?' What's that? Contains carbon?" I would then smile politely, reminding my questioner that I had just completed a major in biology and I understood the term "organic." I used the term, as I do now, to refer to a set of farming practices focused on using animal and plant residues to increase soil fertility.

Organic agriculture surfaced in farming literature of the 1950s, bolstered by Robert Rodale's work in regenerative agriculture. Prior to that, in the 1940s, Masanobu Fukuoka was developing his ideas of farming without cultivation and chemicals. His landmark work, *The One-Straw Revolution*, greatly increased interest in sustainable farming. Since then, dozens of writers, farmer-authors, and researchers have invoked not only the term "natural" farming but also the term "organic."

The increasing industrialization and "chemicalization" of agriculture, together with rising concerns about energy use in agriculture and pesticide poisoning of non-target (non-pest) organisms, prompted expansion of research in the 1970s on alternative organic systems. I myself was inspired by David Pimentel, an entomologist at Cornell University who based his research on agriculture in both a sustainability and an ecology and systematics framework. He was a terrific mentor and supervised my graduate research on a comparative study of organic and conventional wheat production. Those cases, in New York State and Pennsylvania, featured a range of farms, including old-order Mennonite ones.

My research results (Berardi 1978) gained some recognition, as did work by my colleague, Willie Lockeretz, who, at the time, was studying large-scale mixed grain/livestock operations. He has since published on a variety of organic farming systems, both at the grower and at the policy level, and I have published research on other systems—dairy, tobacco, subsistence fishing—in the United States and in Italy. At the time that we were designing our comparative organic-conventional systems research, there was considerable skepticism and concern regarding our motivations, much less the research implications. As I alluded to above, resistance in my land-grant university was high—I conducted it in the Department of Natural Resources, which was more open to systems research of this kind. Researching in such a supportive environment was key to completing my research.

I continued with my research and also joined the board of *Biological Agriculture and Horticulture* soon after its inception—a UK-based publication (then) with a highly international editorial board. Some of my work on organic farming with Pimentel and Sarah Fast (Pimentel et al. 1983) we called "energy-efficient farming" research. It focused on corn, wheat, potatoes, and apple production; yields varied, but organic farming was consistently rated higher in energy efficiency and labor input.

David Pimentel continued to assess such systems for the rest of his career, comparing, for example, conventional and organic farming successes in climate- and nutrient-stressed settings (Pimentel et al. 2005). Other organic farming researchers included Ton Baars of the University of Kassel. My connection with Baars was through the Natural Sciences Section of the Goetheanum in Dornach, Switzerland, where I have studied with students each summer since 2010. John Reganold also emerged as a research movement leader (1988). Dan Nessly reviews much of the literature on this topic.

Also, noteworthy are the long-term experiments started in the early era of research—dating even to the 19th century—in Rothamsted, England, and in the Morrow Plots in Illinois, which document crop productivity, soil quality, and economics in various farming systems (Delate et al. 2017). Much meaningful and substantive research also has resulted from The Rodale Institute's Farming Systems Trial of organic and conventional grain cropping systems (which, in addition, looks at the effect of tillage vs. no tillage) and the long-term biodynamic, organic, and conventional farming system comparisons by the Research Institute of Organic Agriculture (or FIBL, Forschungsinstitut für biologischen Landbau, Fliessbach et al. 2007). The FIBL experiments support hypotheses about a net nitrogen deficit being common after harvest under many systems, and they document favorable soil biological and chemical changes under various management systems. All of this long-term research is exceptionally important to our understanding of what "organic" really means. Some of it suggests, for example, that maybe just a few elements of organic systems (natural sources of fertilizer, say) might be effective in improving soil quality. Other results suggest that unexpected findings—for example, that even organic systems can deplete the soil of nitrogen—are possible with design and funding of long-term studies. Some of the trials look at the possibility of organic no-till systems—a sustainability conundrum.

Today, published research on innumerable aspects of organic farming systems is readily available, far too much to list here. In particular, data on soil organic matter, carbon cycling, and biological activity abound (Lori et al. 2017; Sheoran et al. 2019). Considering the research, what are the implications for farming practices and policy? We now know, especially with research on the biomantle, that the bioactive surface layer of the soil is especially important—we need to regularly provide good doses of organic matter in order to get the microbial activity and plant growth necessary for a resilient food system. Is this enough research to prioritize organic systems?

A common theme in much of the research on organic farming is that it promotes favorable biological soil properties. Should that be sufficient for legislators to focus on changing food and farm policy to promote organic farming, if the objective is to increase resilience in agriculture? Of note is that organic systems promote resilience, too, with their lower energy inputs (Berardi et al. 2011). Considerable research further suggests that organic farming systems can make a difference in increasing resilience to threats such as climate change, urbanization, and diseases—whether it's human pandemic or endemic plant and livestock diseases. Much research today is focused on whether or not organic farming makes a difference to the ecological integrity of natural systems and the financial solvency of farming systems—but there is also the question of the nutritional quality of the food products themselves.

References

Many references for this article were deleted for purposes of brevity and readability; for those references, see: https://wp.wwu.edu/gigiberardi/2020/04/26/early-research-organic-farming/

Berardi, Gigi. 2020. *FoodWISE: A Whole Systems Guide to Sustainable and Delicious Food Choices*. Berkeley: North Atlantic Books.

Berardi, Gigi Maria. 1978. "Organic and Conventional Wheat Production: Examination of Energy and Economics." *Agro-ecosystems* 4 (3): 367–76.

Berardi, Gigi Maria, Rebekah Green, and Bryant Hammond. 2011. "Stability, Sustainability, and Catastrophe: Applying Resilience Thinking to US Agriculture." *Human Ecology Review* 2011: 115–25.

Delate, Kathleen, Cynthia Cambardella, Craig Chase, and Robert Turnbull. 2017. "A Review of long-term Organic Comparison Trials in the US." In *Sustainable Development of Organic Agriculture*, 101–18. Palm Bay, FL: Apple Academic Press.

Fliessbach, Andreas, Hans-Rudolf Oberholzer, Lucie Gunst, and Paul Mäder. 2007. "Soil Organic Matter and Biological Soil Quality Indicators after 21 Years of Organic and Conventional Farming." *Agriculture, Ecosystems & Environment* 118 (1–4): 273–84.

Lori, Martina, Sarah Symnaczik, Paul Mäder, Gerlinde De Deyn, and Andreas Gattinger. 2017. "Organic Farming Enhances Soil Microbial Abundance and Activity—A Meta-Analysis and Meta-Regression." *PLoS One* 12 (7): e0180442.

Pimentel, David, Gigi Berardi, and Sarah Fast. 1983. "Energy Efficiency of Farming Systems: Organic and Conventional Agriculture" *Agriculture, Ecosystems & Environment* 9 (4): 359–72.

Pimentel, David, Paul Hepperly, James Hanson, David Douds, and Rita Seidel. 2005. "Environmental, Energetic, and Economic Comparisons of Organic and Conventional Farming Systems." *BioScience* 55 (7): 573–82.

Reganold, John P. 1988. "Comparison of Soil Properties as Influenced by Organic and Conventional Farming Systems." *American Journal of Alternative Agriculture* 3 (4): 144–55.

Sheoran, H. S., R. Kakar, N. Kumar, and Seema. 2019. "Impact of Organic and Conventional Farming Practices on Soil Quality: A Global Review." *Applied Ecology and Environmental Research* 17 (1): 951–68.

Gigi Berardi, a professor at Western Washington University (https://wp.wwu.edu/gigiberardi/), *is the author of over 25 academic articles and books and over 300 popular articles and reviews. She, among others, pioneered research on energy and economic aspects of organic vs. conventional farming. Her recent book,* FoodWISE: A Whole Systems Guide to Sustainable and Delicious Food Choices, *has received acclaim and endorsements from institutions such as IFOAM Organics International and OrganicEye, as well as over 30 others.*

How Do I Build Climate Resilience for Our Farm?

Thorsten Arnold

The climate is changing, and farmers are the first who are impacted by the small shifts in seasons. In Southwestern Ontario, we had three horrible seasons—2016 as one of the driest years on record, 2017 one of the wettest, and then there was 2018: a very late snowstorm end of April held back spring and, at the same time, was the last rainfall until August! The spring drought overwhelmed our irrigation system and caused wide germination failure in our region. And when we finally saw the first rain in August, it would not stop raining for weeks, and the fields were inundated with water and weeds, destroying much of our yield. My wife, who is the main farmer at home, and I asked ourselves: Are we willing to deal with even more uncertainty in farming, and if so, how are we setting ourselves up to become more resilient?

Clara Nicholls, the president of the Latin American Scientific Society of Agroecology, defines climate resilience as "the ability of a farming system to absorb climate-related disturbances and adapt to stress and change while retaining its productive structure and yield." A resilient farm is not immune to climate threats—but it manages risks successfully by reducing vulnerability and building adaptive capacity.

In order to build climate resilience, a few questions stand out. The first one: What type of climate adversities are we talking about? Are we setting ourselves up such that we can withstand more variability? How would we recover from a severe event, such as a wind or hailstorm? From breakdown of our marketing venues? Or is it the secondary effects of climate change that are our concern—the changes in society and the erosion of values when global climate change forces two billion humans to leave their homes by 2050? We can choose to be resilient against some climate adversities, but our strategy may not prepare us against all! At our place, we expect that the main adversities are increasing variability, the potential for locally destructive storms, and an increasing randomness of our government's actions that we don't want to depend on.

The second question is about the "who." Who should be resilient against these climate adversities? Is it our crop production? Is it my wife as the main farmer and her emotional, physical, and spiritual health throughout the farming season? Is it the financial viability of our farm and its ability to withstand the tax year? What about our marketing venues? Without sales venues, the best crop is useless. In our case, we aspire after a resilient farm with resilient farmers who are embedded within a resilient value chain! In the long term, we want to offer a sanctuary for humans and offer food, labor opportunity, community, spiritual recuperation, and education for regenerative living.

This framework offers two leverage points: we can minimize the vulnerability of our "who entities" against climate adversities that we believe are relevant, and our reactive capacity can recover after a breakdown. And this is when resiliency planning becomes technical!

To reduce vulnerability, scientists recommend strategies and production practices that increase diversity of species, crops, marketing venues, support networks, and community relationships. And scientists further recommend building soil health, the microbial richness of our soil that benefits our production in so many ways: it absorbs water after hefty rainfall, holds it like a sponge and avoids erosion and downstream flooding, retains moisture for drier days, replenishes the groundwater that we can pump during a drought, enhances plant immunity against pests, and gives plants access to nutrients in the soil.

For us, we identified more strategies: low initial investments into external inputs so that we can abandon a crop without undue financial pain, minimum reliance on heavy machinery so we can access our fields at all times, and earthworks that buffer our fields against many weather extremes. After learning about no-till vegetable production from Singing Frog Farms, we decided to adapt their strategy: reduced tillage and compost to build soil health, the use of landscape fabric for weed management that also enables early planting in wet seasons, transplanting almost everything so we have optimal control over the earlier growth stages of crops and a head start on weeds and faster turnaround, and lots of mulching and staking.

Options to build reactive capacity include alternative incomes, collaboration with other farmers (marketing, equipment sharing, processing, learning, research), public education (youth and adults), advocacy and lobbying—all strategies that increase our ability to recover from climate disturbances. Living in the Snow Belt with 14 inches of snow on average, we shied away from shifting to greenhouse production: we would spend all winter fighting snow. We would rather invest into our community and seek off-farm incomes during our long Ontario winters by building sustainable food chains. This is a good example how shifting to greenhouse production would reduce production vulnerability at the expense of enhancing our reactive capacity! This choice allowed our farm to be instrumental in building Eat Local Grey Bruce, an online farmers' market and farmer cooperative. Eat Local Grey Bruce not only created a resilient marketing and community network but also offered a platform that develops local leadership. Our community will need such leaders to recover after severe climate disruption and give a voice to our citizenship.

This year, our investments worked out—we harvested much earlier and nicer than ever before, while other farms were struggling with a late and wet spring. Yet there is no blueprint to building climate resilience: It all depends on where you live, who you are, and how you farm. But we would not recommend farming without being clear on your resiliency strategy.

Thorsten Arnold co-owns Persephone Market Garden. He also works as food chain consultant, offers workshops on climate resiliency planning, and runs a course called "The Barn Academy—From Eco Grief to Becoming an Agent of Change."

By Far, the Most Difficult Real Estate to Transition Lies between the Ears of the Farmer
Dave Bishop

Over the years, I have mentored many (mostly young) farmers transitioning to organic, sometimes informally, but more often through formal mentorship programs such as the Midwest Organic and Sustainable Education Service (MOSES) that offers broad-based support programs to both established and beginning farmers transitioning to organic. In many cases, these are also generational transfers, young men or women taking over their family's farms.

One of my mentees, John, is a young, just-out-of-college farmer planning to transition 2000 acres of conventional farmland owned by multiple family members, not all of whom are confident that this organic thing is a good idea. While strongly motivated to keep the land in family hands, they struggle with how much risk they are willing to accept, especially in inexperienced hands. After much discussion, John got 20 acres to prove himself. "Show us what you can do!" Mistakes made on 20 acres are a lot less expensive than mistakes made on 2000.

Others are forming relationships with neighbors to share machinery and labor, much the way farming communities worked in my childhood. One farmer I work with, who is transitioning 3500 acres of cropland, is working with a neighbor who has cattle and needs additional grazing land. Transition is not just about how we farm the land, but about how we create the kind of community we can thrive in.

Other mentees are older farmers fed up with shrinking profit margins and seeing no other way to escape the tightening grip of corporate control over the food system. Still others struggle to believe in a system that keeps telling them that we can put poison on our food and not poison ourselves.

Characteristics of Successful Organic Farms

A strong focus on building and maintaining healthy soils with a minimum of purchased inputs is a universal characteristic of successful organic farmers. Cultivating a "feed the soil" rather than a "feed the crop" mentality is an essential first step in the transition process. Techniques such as extended crop rotations, effective use of cover crops (green manures), and livestock inputs (brown manures) provide effective means to supply fertility for immediate crop needs and nutrients and organic matter for long-term soil building. Refining these techniques with time and experience increases productivity and improves the quality, or nutrient density, of the foods produced. And since nutrients are what you taste (remember, food doesn't feed you, nutrients feed you; food is just a delivery mechanism), your products soon stand out in the marketplace.

Successful organic farmers look for multiple ways to keep the land productive all year around, rather than use it half the year to grow one thing. Land costs are the same either way. Cover crops protect the soil from erosion; sequester (and generate) nutrients; build organic matter, reduce weed, pest, and pathogen pressure; break up compaction; and generally improve the tilth and biological health of the soil. Grazing livestock provides additional benefits to both the soil and the farmer's bottom line.

Not everyone agrees, but I believe that a truly regenerative farm cannot be maintained long term without having some combination of plants and animals on the land together. Livestock gives us both economic and agronomic benefits, providing the soil with essential nutrients not available from plant sources and providing the farm economy with additional income streams. For example, in the Midwest, adding a small grain to a corn/soybean rotation not only breaks up pest and disease cycles and reduces weed pressure but also offers summer grazing for cattle or other livestock if a frost-seeded cover crop like red clover is included. A legume combined with grazing or an application of manure from winter feeding provides fertility for a coming corn crop and, from our experience, up to 500 dollars of additional income per acre. A high degree of well thought out diversity is the best crop insurance available.

Transition is also an excellent opportunity to redesign the farm's production system to add value to the raw materials being produced. The USDA estimates the farm share of the U.S. food dollar at about 15 cents, with the remaining 85 cents generated by the value adding and marketing of the raw materials produced on the farm. On-farm or community-based processing and distribution assets add to farm income, create jobs, and keep wealth in the community.

The local and organic food movement today offers farmers the opportunity to generate substantial premiums for growing specialty crops that can be sold to local processors or processed on farm and sold directly to consumers. Feeding livestock provides a means to add value to grains grown on the farm. Farmers who find ways to create more value from the land assets they already have will be better positioned to withstand the stresses of transition, and they will go on to build successful organic operations.

Increasing the Chances for Success

The chances for a successful transition to organic can be greatly enhanced, and be less traumatic, if the farmer is willing to extend the process at least a couple years beyond the minimum required 36 months. This gives the farmer, as well as family members, landlords, and lenders, more time to think through the process, and it gives soil ecosystems accustomed to regular doses of water-soluble fertilizers and pesticides more time to recover from chemical dependence.

An extended plan might begin with doing an extensive, detailed soil and tissue analysis and then correcting the deficiencies and imbalances. A couple of years of raising non-GMO crops, experimenting with cover crops and non-chemical weed control techniques such as using a rotary hoe and cultivator while still having the option for a conventional "rescue," lowers the initial stress level significantly. And it's not just for the farmer, but for all those who have a stake in the farm's success.

Establishing a relationship with a certifier as the process begins increases the comfort level of both parties and gives the farmer a little more "get acquainted" time with the paperwork.

Organic farming is a long-term commitment to care for the land and feed the community around us. There is no greater honor than to be entrusted with the care and keeping of a piece of the earth. Use it gratefully!

References

https://www.ers.usda.gov/data-products/food-dollar-series/documentation/#marketing.

Dave Bishop and his family own and operate PrairiErth Farm, a 480-acre diversified farm in central Illinois. They produce corn, soybeans, wheat, vegetables, beef, pork, eggs, and honey. The farm has been certified organic since 2004.

The farm's mix of livestock and crops is the foundation of its sustainable system. Dave's extended crop rotation includes grazing time for livestock on row crop fields to build organic matter in the soil, provide balanced fertility for future crops, and increase the income-producing capacity of each acre.

The Bishops' farming practices have garnered numerous awards: They are the 2017 MOSES Organic Farmers of the Year and recipients of the 2018 RJ Vollmer Award for Sustainable Agriculture from the Illinois Department of Agriculture. In 2019, PrairiErth Farm was one of seven U.S. farms recognized by "Good Food 100" Restaurants and the James Beard Foundation for their dedication to sustainability, transparency, and advancing good food.

Implementation of Organic Enterprises on a Mid-Scale Farm

Chris Bardenhagen

Introduction

In this "Perspectives" essay, I write about my ongoing experience with organic production on our 80-acre farm in northern Michigan. The farm has historically produced fruit and vegetables, mostly potatoes, cherries, and apples, but over the last 15 years, we have incorporated various organic enterprises. My father and I worked together to transition our tart cherries to certified organic, and for several years, I produced organic hay, small grains, and some organic broiler chickens. We therefore have had a "mixed" farm, growing both conventional and organic crops, and that mix has evolved over time. The first part of this essay will focus on our implementing organic tart cherry production and related issues, and the second part will focus on our current planning for new organic enterprises, now that our 20 acres of tart cherries are being removed due to age.

Implementation of Organic for Tart Cherries

In 2008, organic farming was generally booming, and studies were coming out that documented its beneficial effects on farm soil and economics, especially for field crop systems. Upon hearing that there was a significant market for organic tart cherries, my father and I decided to give it a try. However, organic tree fruit production was still on the frontier; fruit farmers in the Midwest were just beginning to successfully produce organically. The transition was a great challenge, but we met it and learned much in the process.

We learned that the high cost of initial inputs is an important consideration. We began to work with an organic consultant to determine what fertilizers and other inputs to use based on soil tests from our orchards. While we were able to purchase composts, fish oil emulsions, and mined minerals such as lime, many of these were both difficult and expensive to source. We were fortunate to find a compost maker that carried several important organic input products located about 100 miles away, though a trip to pick up these inputs would take up most of a day.

We also learned that many organic inputs build soil over time (3–10 years) but don't necessarily feed the crop directly the way conventional fertilizers do. This meant that we experienced lower yields during the transition period, while at the same time, our cherries had to be sold in conventional markets because they were not yet certified organic. This financial "double-whammy" of higher costs and lower yields is important to plan for when transitioning to organic. Some farmers transition only a certain percentage of their crops at a time, in order to keep their farm's income sufficient during transition years. In our case, we were able to draw income off our conventional sweet cherries and apples during the transition period.

The good news began to occur after two or three years, when we really began to notice changes in soil and in the fruit. The soil microbes, uninterrupted by weed sprays and short-term fertilizers, became abundant, and one could literally smell that the dirt had changed for the better. While the cherries were a bit smaller than average, they were noticeably sweeter to the taste and had a higher brix count. Everybody who tried our tarts noted the difference. After three seasons, our fruit was marketable as certified organic, and we were able to get a nice premium for the cherries for several years.

Unfortunately, several years later, an infamous fruit pest (spotted wing drosophila) began to infiltrate our region. This caused a high risk of crop loss due to the lack of available methods to manage that pest. After investing large amounts of money, time, and "TLC," we had to make the very hard decision to take our trees out of organic certification. This was saddening as I very much enjoyed organic production; it was a great fit for my values of cultivating the soil, focusing on long-term sustainability, and raising high-quality products having high flavor and nutrient levels.

Restart into Organic

Click forward a half a decade, and our tart cherry trees are now at the age where they need to be removed. This creates a great opportunity, because I can use those acres to raise other products organically, such as small grains and hay. One benefit of working with these types of organic crops is that their farming practices are much more well tested and researched than with organic tree fruit, and there is a greater availability of farmers with existing know-how and resources such as seed.

At the time of this writing, I am planning the crops and rotations that will go in the space that the tart cherries were in. Items to consider are: What is the length of rotation between crops needed to promote soil health and pest and disease resistance? Which grains are most marketable, and who are the buyers? What are the trucking and storage costs? Also, because we have a mixed farm, when mapping out fields I have to leave about 30 feet between conventional and organic plots.

Can I grow hay on these buffer strips and use it for mulching our conventional apple and sweet cherry trees, a practice that would provide benefits to the health of those soils?

Tentatively, my plan is to use a three-year rotation consisting of oats for year one, clover to provide nitrogen and hay for year two, and hard red spring wheat for year three. There are several organic millers in Michigan who might be interested in my spring wheat, and I have the capability to truck the product to them. Having a wheat-oriented production system would also hold open future opportunities for adding value by starting a small organic milling business in our area.

Moving Forward

Organic farming has its challenges, especially for particular crops (e.g., tree fruit in the Midwest.) It is a slow process to transition conventional soils into productive organic soils, and it requires a high level of investment using both money and time. However, it is exciting and fulfilling to see the soil begin to develop into a more vibrant, living system that supports the needs of crop. I am excited to get back into organic farming with some new enterprises after my brief pause.

Chris Bardenhagen is a PhD candidate at Michigan State University. His research is focused on collective agricultural business organization, with an interest in the incorporation of sustainable practices into production standards. Coming from a farming background, Chris has engaged in both conventional and organic production of fruits, grains, and livestock.

Future Small Organic Farmers Can Be Successful, but We Need Better Policy Support—We Can't Make a Living

Adrian White

Young people like myself want to farm organically. But every year, at least one young farmer in our community here in eastern Iowa drops out of the business. Some of us just can't seem to make a living.

My partner and I—organic farmers for 10 years, with 20 years' combined experience—are in our third season running our very own farm for the first time, Jupiter Ridge LLC. In late 2016, we finally found land to start this enterprise. Before this, we interned, worked at, and managed other organic farms all over the United States. If anything stuck with me all those years (more than learning perfect broccoli spacing or how to milk a goat) was you don't make money farming organically on a small scale. One person, whether among a family, group, or pair of low-income people, must get a job.

My husband and I luckily haven't ever loved organic farming for the money. We love it for the public service, providing healthy food for communities (and the planet), the lifestyle, being our own bosses, the proximity to nature, the passion of just doing it. We continue to love it. We laid the groundwork for the financial realities far ahead of time: like buying or renting land, building infrastructure, purchasing tractors or other equipment, and so on. I even started a successful writing career to prepare us for the extra income needed for start-up. We were fully prepared for what we could face.

Three years later, our farm profits from restaurants, farmers markets, and CSA subscribers are doing well. We've made more money this year than years before. People love our diverse vegetables, mushrooms, and herbs; word has spread because we market ourselves well. We've minimized debt and, in our second year, operated in the black. Last year, my writing made nearly $60,000, all going toward bills and expenses for start-up. On paper, we're running a great, well-strategized business. But we're still not making enough money to cover our costs yet.

Last year, our farm only made us $1000. We're still slowly investing in costly (but required) start-up to avoid debt, something we can only do over the long term because we aren't wealthy. Expanding that tiny profit margin feels a long way out. Though we operate with the same challenges as conventional farmers, we don't use the environmentally harmful shortcuts and government subsidies that make conventional farming livelihood "more" viable (though some small family conventional farms still struggle in similar ways, too).

Everything in our society, as it's set up now, feels dead set against new small organic farmers making a living. This needs to change. Unless you have prior wealth, exploit workers for low pay, inherit an all-included farm business, or go "big ag" mechanized organic, getting things off the ground is an uphill battle—and wealthy or no, this just won't generate the future farmer population we need.

Compared to conventional production, organic production costs tend to be higher and are much more labor-intensive (Post and Schahczenski 2012). Harmful conventional pesticides and herbicides are designed to cut down production costs, but at a steep environmental price. Instead, we organic farmers must compensate with more time and labor (manual, mechanical weeding, or employee labor), higher product prices, and—in some cases—organic pesticides or herbicides (which tend to be more expensive than conventional ones). All this adds up production costs, and as we all know, time is money when it comes to labor.

To farm, you also need land. The average per-acre price of farmland is around $3000, according to the USDA's 2019 Land Values summary (USDA 2019). It gets more expensive every year. Even if a hopeful organic farmer finally musters up funds to buy or rent land, they still have high-priced infrastructure, equipment, and operating costs to reckon with on top of the steep production costs.

It's a regular sentiment among organic farmers, including myself, to feel on the brink of quitting sometimes with these financial challenges. Some days they feel very close; other days we forget this knife's edge. Sometimes I look at neighboring fields of corn and soy and realize that if crop failure or weather were to strike, those conventional farmers would probably get a subsidy check. For what we grow, we'd get nothing.

And sure—the 2018 Farm Bill advanced funding for organic agriculture (O'Neil 2018). But these provisions appeared to be aimed at big ag operations, incentivizing them to transition to organic. Something like crop insurance, subsidies, or even student loan forgiveness for new organic ag entrepreneurs would be a better idea. It would finally acknowledge farmers as the public service workers they've always been.

But it's not all bad. That's my biggest message. Young organic farmers do hold the ultimate impassioned public service jobs of our day. They're not categorized as such, but they should be. Despite no incentives or millionaire potential, organic farmers sequester carbon through regenerative agricultural practices, which could be huge for climate change if done on a mass scale. They grow whole, healthy foods that, with help from government programs, could improve public health and nourish the poor. They trade stale office jobs for the outdoors and nature. For me, I still can't imagine myself doing anything else besides this—the real struggle I face is the fear of *not* being able to do this for much longer.

I encourage all who care about today's greatest issues—poverty, inequality, climate change—to include organic farming as a central solution to all those things. Until policies change to better support new organic farmers, especially reducing their start-up burdens, we'll keep experiencing the slow loss of these public service workers we have. If not, we prevent a whole willing and passionate generation of people from solving today's biggest crises.

In the meantime, I'll keep watching as more and more young organic farmers drop out. I'll keep hoping I can continue doing what I love—even if they weren't able to.

References

O'Neil, Colin. 2018. "The 2018 Farm Bill Is a Big Win for Organic Farming." EWG.com, AgMag. https://www.ewg.org/agmag/2018/12/2018-farm-bill-big-win-organic-food.

Post, Emily, and Jeff Schahczenski. 2012. "Understanding Organic Pricing and Costs of Production." National Sustainable Agriculture Information Service (ATTRA).

United States Department of Agriculture (USDA). 2019. "Land Values 2019 Summary." National Agricultural Statistics Service.

Adrian White is a writer, organic farmer, and herbalist. She is co-owner and operator of Driftless Iowa organic farm and Jupiter Ridge LLC and the owner of Deer Nation Herbs. Her writing work can be found in The Guardian, Civil Eats, Good Housekeeping, *and various other publications all over the web, as well as on her website,* IowaHerbalist.com.

CCOF
Certified
Organic
Certified
CCOF
Organic
www.ccof.org
Certified organic by CCOF in accordance with the USDA National Organic Program.
Organic Certification, Trade Association, Education & Outreach, Political Advocacy

Introduction

The organizations and individuals profiled in this chapter represent a very small fraction of those who have contributed to the organic movement. The individuals profiled here are considered the founders of the movement or have contributed significant policy development, knowledge, and support to the organic movement in the United States. The organizations profiled are those who have a strong commitment to organic agriculture and have made noteworthy contributions to organic research, advocacy, policy, or awareness. There are many more organizations and individuals that are dedicated to supporting the organic movement, some of which can be found in the resources chapter of this book.

People

Lady Eve Balfour (1898–1990)

Lady Eve Balfour is considered a pioneer of organic farming and was certainly one of the most vocal and visible proponents of the early movement. Lady Evelyn Barbara Balfour, the fourth of six children, was the daughter of Gerald Balfour, the Second Earl of Balfour, and Lady Elizabeth Balfour. Despite growing up in a political family, she had an interest in farming

---

Hall Wines in Napa Valley, California, is certified organic by the CCOF. They are one of the oldest organic certifiers on the West Coast. (Wollertz/Dreamstime.com)

at a young age. Despite a lack of direct farming experience, she was one of the first women to attend Reading University College from 1915 to 1918. She spent two years studying at the leading agricultural college and did a year of apprenticeship at Manor Farm (the experimental farm connected to the university). Very few women studied farming at the time, and she was the only female apprentice on the farm. Lady Balfour's first job in 1918 was to manage a small farm for the Monmouthshire War Agricultural Executive Committee.

In 1919, Lady Balfour purchased a 157-acre farm called New Bells with her sister Mary in Haughley Green, Suffolk. She experimented with modern farming technologies at this time, including many of the new synthetic fertilizers. In the 1930s, Balfour and her sister added a market garden to their mixed farm and opened a farm shop in London. Over time, she expanded the farm, buying up a neighboring farm that fell during the Depression. She also took up political work fighting against tithes to the church and writing articles about the impacts of free trade on small farmers. Her interest in health and nutrition paired with an interest in supporting local farmers was a precursor to her interest in organic production methods.

In 1938, Lady Balfour read a book by Gerald Wallop titled *Famine in England*, which introduced her to the concept of organic farming and convinced her to explore the methods on her own farm. The book drew on work by Sir Albert Howard, Robert McCarrison, GT Wrench, and Ehrenfired Pfeiffer, which espoused a biological view of the soil and encouraged farmers to forgo chemical fertilizers in favor of composting to build up soil fertility. The book suggested there was a strong connection between soil health and human health. Shortly after reading the book, Eve transformed her two farms into a research site aimed at comparing organic and nonorganic farming systems with the goal of proving organic agriculture produced healthier food. This became known as the Haughley Experiment. Lady Balfour and her farming partners set up her two farms with side-by-side plots of organic and conventional production practices. In this way, they were able to compare organic and conventional or chemical-based growing techniques on the same soil and in the same weather conditions. The goal was to prove that increasing soil fertility through chemical-free composting-based methods would result in healthier soil, animals, and thus people. The experiment was the first to do a side-by-side comparison with the goal of long-term study. The Haughley Experiment was taken over and run by the Soil Association from 1947 to 1969.

In 1942, Eve wrote a book, *The Living Soil*, that became a foundational text for the organic movement. The book, published in 1943, was a bestseller that actually started as a memorandum for the Haughley Experiment fundraising efforts. In the book, she presents evidence of the importance of soil to human health and that for soil to be fertile it needs to be biologically alive. She includes instructions on how farmers and gardeners can build soil fertility through composting. The book went through many reprints into the 1970s and is still widely referenced today. In 1944, the BBC commissioned a series of talks based on *The Living Soil* for the program *Radio Trek*.

Following the success of the book, Balfour and others were encouraged to cofound the Soil Association in 1946. She was a vigorous campaigner for the organic movement and held a wide variety of positions within the Soil Association until she retired in 1984. She traveled widely, wrote for numerous publications, and lectured in support of organic farming for many years. During this time, she was invited to be a founding member of the International Federation of Organic Agriculture Movements. One of her most notable lectures was at the IFOAM international meeting in 1977 titled "Towards a Sustainable Agriculture—The Living Soil." For her work in the organic agriculture movement, she was made an Officer of the Order of the British Empire in 1990, shortly before she died. In a fitting tribute, the day after she died the British government announced grants to support farmers converting to organic agriculture.

Eliot Coleman (1938–)

Eliot Coleman is an organic vegetable farmer who has influenced many generations of organic farmers with his organic gardening books that describe innovative production methods for small-scale market growers. Eliot was born and raised in New Jersey, with no farming background. He attended Williams College, where he studied geology and Spanish literature. Eliot loved the outdoors and was intrigued with the idea of earning a living on the land. He first started learning about farming and gardening by reading old farming books from the early 1800s.

Eliot and his first wife, Sue, moved to a 60-acre farm in rural Maine in 1968 to live a self-sufficient life. The couple had met in 1964, at Franconia College in New Hampshire, where Eliot was teaching Spanish, and they married two years later. They had previously read *Living the Good Life: How to Live Simple and Sanely in a Troubled World*, by Helen and Scott Nearing, who encouraged young people interested in their way of life to come visit them and learn their ways. After visiting the Nearings and wanting to stay, Eliot and his wife bought a plot of land from them for $2000 and became part of the "back to the land movement" of the 1960s and 1970s. They built their own home and created a self-sufficient farm from the forested land.

During their early years on the farm, their lifestyle was profiled in the *Wall Street Journal* and shared among many people interested in the growing trend of homesteading and organic farming. They had many young people come to stay at the farm to learn how to raise their own food and live sustainably. After a tragic family accident in 1976, the family left the farm for a time. Eliot, though, continued to pursue his passion for organic farming, including writing about it for a number of publications. He was an active member of the Maine Organic Farming and Gardening Association, speaking and writing for their events. To augment his learning, Eliot spent considerable time traveling across Europe to learn new techniques that could be incorporated with his own.

He eventually returned to the farm in Harbourside, MA. Along with his second wife, Barbara Damrosch, a well-known vegetable gardener and writer, they resurrected the farm as Four Season Farm, where they continue to sell organically grown vegetables direct to consumers year-round. The farm has become very successful, bringing in six-figures with just a small amount of land under cultivation. They sell locally, aiming to serve those within 40 miles of the farm. They sell to restaurants and stores and at farmers markets and their roadside stand. Eliot continued to educate the next generation of organic farmers and gardeners through apprenticeships, lectures, and writing.

In 1989, Eliot published his first book, *The New Organic Grower*. A second edition was published in 1995, and a 30th anniversary edition was published in 2018. The book has been a foundational text for many organic vegetable farmers. He wrote the *Four Season Harvest,* published in 1992 and 1999 to document the techniques he uses to grow food year-round in a northern climate. His use of cold frames, plastic-covered greenhouses, and succession planting are explained with detailed instructions and complemented with his philosophies on growing food year-round. He followed this book up with *The Winter Harvest Handbook*, which builds on the techniques introduced in the *Four Season Harvest.* In this book, he goes into detail about all aspects of running a market garden from planning to planting, harvesting, and even marketing strategies. Along with his second wife, Barbara, he co-hosted a TV series called *Gardening Naturally*, which aired on The Learning Channel. And they co-wrote a cookbook, *The Four-Season Farm Gardener's Cookbook*, published in 2013.

Both hobby gardeners and small-scale commercial growers rely on the strategies he developed for growing a diverse array of crops and making a profit on a small scale. His books are unique in that he goes beyond production techniques and includes detailed instructions and drawings for building structures and tools that make his techniques so successful. He expanded his love of developing new tools by designing and consulting for a number of companies, including Johnny's Selected Seeds.

Eliot is well known not only for his practical instruction but also for his strong beliefs in the fundamental aspects of organic farming. Eliot served as the executive director of IFOAM from 1978 to 1980. He was also a key advisor to the USDA on their landmark study *Report and Recommendations on Organic Farming* published in 1980. But once the OFPA was passed, he lost faith in government's involvement in the organic farming sector. He has never been certified organic, and he advocates for alternative approaches such as local and small scale. He wrote an article titled "Beyond Organic," which argued that the USDA now owned the word organic, and so it no longer represented the values of the organic farming community, and suggested farmers use the word authentic instead. Currently, Coleman is serving as an advisor with the Real Organic Project, a farmer-led organization dedicated to taking back the meaning of organic by creating a certification system that goes beyond the USDA organic certification. Eliot has been a key supporter of rallies and lobby efforts to protect the integrity of organic on a number of issues. In 2015, Eliot received a James Beard Foundation Leadership Award in recognition of the work he has done for the organic farming sector.

The Howards: Albert, Gabrielle, and Louise

Albert Howard (1873–1947)

Born in Shropshire to a farming family, he studied natural sciences and agriculture at the University of Cambridge. He lectured at Harrison College in Barbados before working as a mycologist and lecturer in the West Indies. His work of teaching schoolmasters required him to expand beyond his narrow research interest to understand the whole process of growing crops through harvest and market. This forced Howard to consider how the work in the laboratory was connected to the work of growing food for markets, and he realized that researchers needed to experience both. Albert returned to England to work as a botanist at the South East Agriculture College in Wye, where he spent time researching and sharing his research directly with farmers. In 1905, he married Gabrielle Howard, and together they moved to India where they became Imperial Economic Botanists to the government of India.

In their early work at Indore, they grew cotton, sugar cane, and cereals, among many others. They grew crops that the locals raised using methods comparable to local practices. Albert and his wife Gabrielle were actually not averse to synthetic pesticides and fertilizers as many people believe. They just did not see them as accessible for the vast population of Indian farmers. They also did not derive their process from the local Indian peasant farmers as Louise, Howard's second wife, wrote in her biography of Albert. These ideas were added to their writing afterward as their thoughts on the issues evolved. The Indore method was developed as a way for local farmers to improve soil fertility, and therefore yield, with an economical approach. They observed that the local soils lacked organic matter and nitrogen, so they developed systems of composting using waste from the farm. The results were increased yields and a decrease in insects and diseases in their crops. They also saw improved health in their farm animals. All waste elements from the farm were incorporated, including manure, crop residues, forest debris, and urine soaked earth from the cattle barn. The details of the Indore Method were published in the book *The Waste Products of Agriculture: Their Utilization as Hummus* in 1931.

After Gabrielle's death in 1930, Albert retired early and traveled to Africa where he saw numerous examples of his composting methods being used. This gave him the idea to expand the reach of his methods. A year later, he married Gabrielle's younger sister Louise, returned to England, and dedicated the rest of his life to bringing attention to the importance of soil fertility. In 1934, Albert Howard was knighted for his wide-ranging contributions to agriculture.

In 1940, Albert published *An Agricultural Testament*, a book written for a popular audience that brought together all his scientific research with ideas he had formulated after retirement. In this book, he describes the basic principles of organic farming, using nature's methods of soil management. He coined the term "Law of Return," which describes the recycling of organic waste material, including human sewage sludge, to return nutrients back to the soil. Albert corresponded with people from around the world, trying to convince people to take up organic farming methods. He had some success in Britain and parts of Africa and Asia, but it wasn't until he started writing for J.I Rodale that his ideas gained popularity in the United States. In 1945, he published *The Soil and Health* (originally *Farming and Gardening for Health or Disease*) with Louise, as a summary of his life's work and the impact it had around the world. In this book, he criticizes the reductionist approach to science and the lack of real-life problem-solving strategies. In this work, he also introduces his ideas about organic farming and intentionally used the term "organic." He relates the idea of growing within an ecosystem and especially the role of forestry in maintaining healthy soils. He further discusses new ideas about how soil fertility and composting could impact the health of both animals and humans. Many of these ideas were gleaned from his interactions with Robert McCarrison, a physician who studied the effects of soil fertility and human nutrition. Throughout the entirety of his work, one of the main constants was the belief that the science of farming must take context into consideration and not succumb to reductionist analyses. Many consider Howard to be the father of organic farming and his books to be foundational to the science of organic agriculture.

Gabrielle Matthaei Howard (1876–1930)

Born in Kensington, England, to a merchant family, she studied at Newnham College in Cambridge and began working as botanist and plant physiologist. She conducted experiments on temperature in photosynthesis. In 1905, she married Albert Howard and worked with him as an economic botanist for the government of India. She was a rare woman of science for her time and was lucky to find a marriage partnership that allowed her to continue to work as a scientist. Together they ran a number of research stations in India, culminating with the development of the Institute of Plant Industry at Indore. They spent 25 years researching together and co-wrote over 120 articles. Gabrielle presented her findings at a number of conferences held at Indore. Gabrielle died in 1930 from cancer.

Louise Matthaei Howard (1880–1969)

Sister of Gabrielle, Louise also attended Newnham College in Cambridge where she studied classics. She worked as a lecturer and editorial assistant before taking a job with the International Labour Organization in Geneva. There she worked her way up to the position of Chief of the Agricultural Service. In 1931, she married Albert Howard, her sister's widower. She brought her extensive knowledge and writing skills to the partnership to support Albert in his work in popularizing the importance of soil fertility. Together with Sir Albert, she published *Farming and Gardening for Health or Disease* in 1945. After her husband's death, she founded the Albert Howard Foundation and published a book called *The Earth's Green Carpet* in 1947, a brief overview of her husband and sister's work in an accessible non-technical format. The foundation later merged with the Soil Association. For many years, she kept Albert's efforts alive through a newsletter called Albert Howard News Sheet, which connected his work to the broader environmental issues that emerged in the 1950s and 1960s. In 1953, Louise published a biography of sorts on Albert's work in India called *Sir Albert Howard in India*, which gave an overview of the main accomplishments during his research years.

Paul Keene (1910–2005)

Paul Keene, along with his wife, Betty, was one of the earliest organic farmers in North America. Paul was born in Lititz, Pennsylvania, on October 12, 1910. He graduated from Lebanon Valley College in 1932 and went on to earn a master's degree in mathematics from Yale University in 1936. From there, he taught math at Drew University in New Jersey. In 1938, after a few years of teaching at the university, he wanted to experience something new and so decided to try teaching at an international school in northern India. While he was teaching in India, he became involved in the peace movement, during which he met Mohandas Gandhi and was inspired by his belief in living simply. Paul also discovered the work of Albert Howard, who had completed years of organic farming research in India just a decade before. During Paul's time in India, he met and married his wife, Enid Betty Morgan, the daughter of missionaries.

Shortly after they married in 1940, Betty and Paul moved back to the United States and spent four years studying organic farming and homesteading. They started at the School of Living in New York state. The School of Living was founded by Ralph Borsodi, an advocate of simple living and self-sufficiency. They worked and taught at the school in exchange for room and board and modest pay. It was there they first discovered the work of Rudolf Steiner and biodynamic farming. In 1943, they moved to Kimberton Farm School outside of Philadelphia to study with Dr. Ehrenfried Pfeiffer, who taught biodynamic farming principles. Again, they exchanged work for room, board, and a small monthly sum. It was at Kimberton Farm School that they became friends with J.I. Rodale, another organic pioneer.

In 1945, they decided to rent their own farm and a year later to buy one of their own. They borrowed $5000 to buy a 100-acre farm in Pennsylvania named Walnut Acres. They had very little money or supplies and two young daughters, but they were eager to make it work. They started off with no plumbing or electricity and only a team of farm horses, a plow, and a harrow. Their first products were eggs and apple butter made in their own kitchen on top of a cast-iron stove and sold for a dollar a jar. Clementine Paddleford, a food editor at the *New York Herald Tribune*, gave them a huge boost in their first year of business by giving them a positive review on the apple butter. Unexpectedly, people started traveling from the city to their farm to pick up their famous apple butter.

From there, they started a mail-order company that sold all sorts of foods. In the early days, they shipped frozen chickens in the winter, eggs in metal containers, and, of course, their famous apple butter. Over time, the business grew to include 600 acres, a manufacturing plant, and a mail-order catalog shipped to over 40,000 customers a year and over five million dollars in revenue. The growth was slow and followed a natural progression. Their ultimate goal was to not waste what they produced on the farm by adding value in the form of processed foods and vice versa. Of the 600 acres, 360 acres were tillable and used to raise organic grains and vegetables that were sent to the on-site manufacturing plant. The factory included a flour mill, cannery, bakery, processing room, and retail store. Remnants from the milling process were sent to the cattle and poultry operation several miles away to serve as feed for the animals.

During the 1980s, Paul's and Betty's health declined, and one of their daughters and a son-in-law took over the operations. Their children made significant changes in the catalog offerings, including imported items and kitchen gadgets and pulling many of the products made on farm. Many of the original customers missed their favorite products. By the early 1990s, only 40 percent of the food that was processed and packaged on-site was also grown on the farm. They expanded to include over 350 products in their catalog. By this time, much of the operation needed to be upgraded. In 1999, in order to raise capital to grow the business they sold a controlling interest to David Cole for $4 million. In 2000, he closed the plant and catalog and discontinued all but 20 products. He then sold the

Walnut Acres brand to Hain Celestial Group in 2003. Many people who had supported Walnut Acres over the years were disappointed in the demise of the company built by Paul and Betty Keene and felt the loss of their strong values in guiding the operation.

In 1998, Paul published a collection of essays about farming with Globe Pequot Press, called *Fear Not to Sow Because of the Birds*. The essays were drawn from his regular columns published in the catalog. That same year, he received the Organic Trade Associations Organic Leadership Award. Over the years, their integrated approach to farming, manufacturing, and sales inspired a new set of organic farmers trying out the same model. Gene Kahn of Cascadian Farms credits Walnut Acres and the catalog with Paul Keene's column as inspiration for his business model. Though their legacy continues in memory only, Betty Keene died in 1987 and Paul developed severe Alzheimer's before passing away at the age of 94 in 2005.

Kathleen Merrigan (1959–)

Kathleen Ann Merrigan was the primary author of the OFPA and continued to advocate for the NOP in a number of roles, including as the Deputy Secretary of Agriculture under the Obama administration. She was born in Pittsfield, Massachusetts, and raised in nearby Greenfield. She attended Williams College in Williamstown, MA, where she received a Bachelor of Arts in English and Political Science. She went on to earn a Master of Public Affairs degree from Lyndon B. Johnson School of Public Affairs at the University of Texas-Austin. Merrigan continued her education at the Massachusetts Institute of Technology, where she graduated with a PhD in Environmental Planning and Policy.

Kathleen began working as support staff to state legislators in Massachusetts and Texas in the 1980s. From 1986 to 1987, she was a special assistant to the chief of regulatory affairs for the Texas Department of Agriculture tasked with working on pesticide issues. From there, she joined the staff of the U.S. Senate

Committee on Agriculture, Nutrition, and Forestry. She stayed in her role as the chief science and technology advisor to the chairman Patrick Leahy, a democratic senator from Vermont from 1987 to 1992. It was during her time as a staffer for Patrick Leahy that she drafted the proposed OFPA. She conducted extensive consultations with members of the organic sector and consumer organizations who wanted strict organic regulations, and she tried to work it all into a bill that would actually pass. Unlike earlier attempts to pass an organic agriculture bill, this one succeeded.

Kathleen continued her work in sustainable agriculture as a senior analyst at the Henry A. Wallace Institute for Alternative Agriculture. During this time, she also served on the NOSB from 1995 to 1999. She left those positions when she was appointed the administrator of the USDA's AMS in 1999. While working in the USDA, she continued to support the organic program and was critical in helping to move along the creation of the NOP. Beyond her work with the USDA, she was involved in the UN Food and Agriculture Organization (FAO) in a number of roles, including as a delegate and the first woman to chair the Ministerial Conference. She was also a professor at the Friedman School of Nutrition Science and Policy at Tufts University for a time.

She was appointed U.S. Deputy Secretary of Agriculture by President Obama in 2009 and served until she resigned in 2013. Her reputation as an organic advocate gave many in the organic sector hope that she would promote and support the sector from within the USDA. That same reputation also gave Republicans concern during her confirmation hearing. She acknowledged their concerns and convinced them that she was interested in supporting all farmers. In her role, she did continue to support and elevate the NOP, but she managed to balance that support with the need to serve all farmers. During her time as deputy secretary, she was also the chief operating officer of the USDA and worked closely with First Lady Michelle Obama to create the Know Your Farmer, Know Your Food initiative.

Although many feel she could have done more for the organic sector, she did take a number of steps to improve the NOP. One of the first steps she took was to initiate an outside audit of the NOP conducted by the National Institute of Standards and Technology within the Department of Commerce. She created an organic program coordinator position tasked with integrating organic into all aspects of the USDA, including requiring all USDA employees take two organic courses. They also implemented the Organic Literacy program, which makes information available to all the USDA field staff on how to help organic farmers. Under her guidance, the USDA set a goal of increasing certified operations in the United States by 25 percent by 2015. She also hired Miles McEvoy, who ran the Washington State organic program for 20 years, to head the NOP. Throughout the years that she was deputy secretary, the NOP managed to get through a backlog of organic complaints and to increase enforcement. She dedicated more support to cost-share programs for organic farmers and negotiated better certification equivalency agreements with the EU.

After working in the government for many years, she took on the role of executive director of sustainability and director of the George Washington Food Institute and was a professor of public policy at George Washington University between 2013 and 2018. Kathleen made headline news in 2018 when she declared that food action was occurring outside of Washington and that it is up to the private sector to move issues forward. She took a new position at Arizona State University as the executive director of the Swette Center for Sustainable Food Systems and the Kelly and Brian Swette professor of sustainable food systems in Tempe Arizona. There she has continued to participate in national organic advocacy.

In addition, she became a partner in the venture capital firm, Astanor Ventures, and advisor to S2G Ventures, both focused on ag-tech innovations. She has served on a number of boards including World Agroforestry Centre, Stone Barns Center for Food and Agriculture, FoodCorps, and Center for Climate and Energy Solutions. She served as inaugural co-chair of the AGree Economic and Environmental Risk Coalition. She was named one of *Time* Magazine's 100 Most Influential People in the World in 2010.

Despite working on behalf of all agricultural entities in her government positions, she remained an advocate for organic agriculture and is often known as the "Mother of Organic" in reference to her writing the law that serves as the basis for the NOP. But her support of organic was not always without controversy. She supports large-scale organic production and encourages big corporations to be involved, which proponents suggest has compromised the organic standards. But Merrigan maintains that growth of the organic sector is a good thing, making it more accessible to a broader population. Unlike others in the organic sector, she embraced new agriculture technology and did not shun aspects of conventional agriculture that most in the organic sector strongly oppose, such as GMOs. In fact, she wrote an article arguing against widespread GMO labeling laws, which then garnered rebuttal headlines claiming she has sold out the organic sector.

Lord Northbourne (1896–1982)

Walter Ernest Christopher James was born and raised in Kent County, England. He attended private schools in his youth and served in World War I. After he returned from the war, he studied agriculture at Oxford University, graduating in 1921. After his father died in 1932, he succeeded to the title of Fourth Baron Northbourne and assumed responsibility for the family estate. He served as the chairman of the Kent Agricultural Executive committee and the Wye Agricultural College for many years.

It is likely that Lord Northbourne's combined interests in agriculture, education, and spirituality drew him to the work of Rudolf Steiner and Ehrenfried Pfeiffer. He was at school in Oxford when Steiner presented a conference there, and it is probable that Northbourne would have attended. He met with Ehrenfried a number of times throughout the 1930s to discuss views on agriculture and health. In 1939, Northbourne traveled to Switzerland to meet with him and help organize the first biodynamic conference in England. As a result, Lord Northbourne hosted the Betteshanger Summer School and Conference on biodynamic Farming in July of 1939 on his estate in Kent. During that time, he was in the process of converting his entire estate to organic and biodynamic farming practices, and thus they used his farm as a demonstration for many of the techniques they discussed during the conference.

In 1940, Northbourne published the book *Look to the Land,* a treatise on organic versus chemical farming. He wrote about the concepts of farming and health that were mainstays in the biodynamic community, but he stripped away the spirituality to make the ideas appeal to a broader audience. Lord Northbourne was concerned about the shift to chemical agriculture that had been taking place and warned against the artificial manure industry. In the book, he wrote of the farm as an organic whole that cannot rely on outside inputs. For these concepts, he referenced the work of Pfeiffer and Steiner, who considered the farm as a living organism. While he does draw heavily on the biodynamic system of farming, he also drew on work by others such as Sir Albert Howard. Northbourne was the first to use the term "organic farming" and set it up as an opposite to chemical farming. Northbourne asserts that the biodynamic system is one approach to organic farming, thus setting up organic farming as a broader philosophy with many different ways to execute it. The book contained a number of principles that are now often associated with organic farming, such as the importance of local food, nature protection, something akin to the precautionary principle, and the biological importance of soil to nutrition and health.

The book was published in Britain and Australia, and in the United States, it was published in part in the *Bio-dynamic* journal. The book was referenced heavily by authors around the world working on similar issues. Lady Eve Balfour quoted large portions of the text in her own book *The Living Soil* and went on to create the Soil Association. In the United States, J.I. Rodale adopted the term and began publishing a journal called Organic Farming and Gardening. In Australia, the Australian Organic Farming and Gardening Society was started. The book was reprinted several times in the 1940s and then went out of print until his son had it reprinted in 2003. For that reason, many have forgotten the role that this classic book played in the history of organic agriculture.

Dr. Ehrenfried Pfeiffer (1899–1961)

Ehrenfried Pfeiffer was born in Munich, Germany, to parents who were active in the Anthroposophical movement and were personal acquaintances of Rudolf Steiner. He attended the University of Basle in Switzerland and studied chemistry, botany, and physics. In 1920, he moved to Dornach, Switzerland, and worked on stage lighting, among other things, at Goetheanum, a theatrical center designed by Steiner. He was the first to test some of Steiner's biodynamic preparations. Eventually, be became the manager of the biodynamic research farm, Loveredale, in Domburg, Netherlands. There he tested and conducted experiments derived from Steiner's agriculture lectures. In 1933, he traveled to the United States to lecture to a group of farmers at Threefold Farm in New York. In 1938, he published *Bio-Dynamic Farming and Gardening* in five languages. He lectured and corresponded with a number of people about the growing alternative agriculture movement, and in 1939, he ran a conference in England with Lord Northbourne, called the Betteshanger Summer School and Conference. With the onset of war, Ehrenfried took up an offer from Alaric Myrin to move to Kimberton, Pennsylvania, and create a model biodynamic farm and training program. He helped to found the Biodynamic Farming and Gardening Association. It was there that he met with J.I. Rodale, who was interested in his work and published several articles by Pfeiffer. Eventually, Pfeiffer bought his own farm in Chester, New York, and started developing commercially viable compost starters. Pfeiffer had significant influence on the development of the organic and biodynamic movement, and he trained many of the early organic pioneers.

Nora Pouillon (1943–)

Nora Pouillon created the first certified organic restaurant in the United States and helped develop the organic certification standards for restaurants. Nora was born and raised in Austria during World War II, and then she fled with her family to a farm in the Alps. After the war, she returned to Vienna to go to school, but she continued to spend summers on her grandmother's farm. Her family was quite health conscious and taught her to cook with simple fresh ingredients.

In 1965, she immigrated to the United States with her husband. The move to the United States opened her eyes to the differences between European and North American cuisines. She was shocked to see all the prepared and processed food that had no bearing on seasons or a connection to the people who made it. Because she loved to cook good food, she began to seek out quality ingredients, often found at coops and health food stores. Eventually, she started driving out to farms from her home in Washington, DC, to find better-quality foods. It was through her relationships with these farmers that she learned about organic farming practices. As friends and acquaintances began to notice her skills with food, she slowly developed a casual catering business and offered cooking classes in her home, all emphasizing the importance of local seasonal food sold by organic farmers.

When she separated from her husband in 1976, she needed to find a way to make her own living. Her friend was opening an inn and decided to have Nora open a restaurant inside the inn. As the restaurant attracted more customers, she decided it was time to open her own restaurant. She partnered with the hotel manager, who eventually became her husband, and together they managed to convince enough friends to invest in the restaurant. Restaurant Nora was opened right in Washington, DC, Dupont Circle in 1979.

She was determined to run the restaurant with organic food right from the beginning even though she got plenty of pushback about it. At the time, many people associated organic food with tasteless health foods. Despite the image organic had of being boring health food, the restaurant succeeded in creating a gourmet seasonal menu with fresh local foods, and success followed. Not only did Nora use local organic farmers to supply her own restaurant, but she also leveraged her success to encourage other chefs in DC to use these same local farmers. She also initiated the first farmers only farmers' market in Washington, DC.

In the 1990s, when certified organic became better known she decided it made sense to certify the restaurant as organic. But she discovered that there was no process for certifying a restaurant, and none of the six certifiers operating at the time had ever considered it. Nora worked with Oregon Tilth for over three years to create a set of standards suitable for restaurants. She and her staff spent two years documenting everything they did to ensure her restaurant was organic. In the end, they determined that restaurants could be certified organic if they sourced at least 95 percent of their ingredients from certified organic farms and imported ingredients. Every year, the restaurant tracked proof of organic certification from all their sources and maintained the required paperwork to prove compliance. In addition, they determined that restaurants needed to ensure everything from cleaning supplies to storage and prep areas all met the established organic processing requirements.

Restaurant Nora became the first certified organic restaurant in North America in 1999, just before the U.S. certified organic regulations came into place. Although Nora's efforts have encouraged many more chefs to buy food from local organic farmers, only a few restaurants have followed her footsteps to becoming certified as the effort and cost to maintain certification is high. Nora felt the certification was worth it to prove to her customers that her claims of organic were reliable and that it creates an awareness about sustainable food in general. In 2017, Nora was disappointed that only a dozen or so restaurants had attempted to run a certified organic eatery. In 2015, she published a memoir called *My Organic Life: How a Pioneering Chef Changed the Way We Eat Today*.

Her restaurant remained successful throughout its 38 years of operation, and even had the honor of hosting most presidents of the United States. She opened a second restaurant called City Cafe and later changed it to Asia Nora, striving to use as many organic ingredients as she could find in that restaurant as well. In spite of its popularity, she closed that restaurant in 2007. In 2017, Nora closed her original restaurant and retired after being unable to find a suitable buyer. She announced that she would continue to advocate for organic food and offer consultations to the organic food industry.

Nora expanded her reach by developing recipes for well-known organic companies such as Walnut Acres, Fresh Fields, and Green Circle Organics. She authored a cookbook, *Cooking with Nora*, which was a finalist for the Julia Child Cookbook Award. She has won numerous awards for both her cooking and her work in the organic sector. She was and remains today an advocate for sustainable organic food, especially in the restaurant sector. Nora has been on the board of directors for a number of environmental organizations and supported a long list of environmental, organic, and culinary organizations focusing on sustainable living. She was active in other areas of sustainable foods including fish, foie gras, and others. She worked with a number of other chefs to create the organization Chefs Collaborative in 1993, to encourage sustainable food practices. She founded Blue Circle Foods as a sustainable seafood company to supply fish to chefs, retailers, and distributors.

The Rodales: Jerome Irving, Robert, and Maria

Jerome Irving (J.I.) Rodale (1898–1971)

J.I. Rodale was born in New York and worked as an accountant for a number of years before joining his brother in founding an electrical manufacturing company in 1923, and he married in 1927. Rodale had poor health and so began to read health and fitness books. When he and his brother moved their company to Emmaus, Pennsylvania, in 1930 he founded a publishing company called Rodale Press. Around that same time, he read a book by Albert Howard and developed a keen interest in the connection between healthy soils with healthy food. He bought a farm in 1940 and began experimenting with organic farming. He started publishing the *Organic Farming and Gardening* magazine in 1942. He invited Howard to write for his magazine and so introduced Howard's composting methods and organic farming principles to an American audience. He met with other organic pioneers such as Ehrenfried Pfeiffer and often invited them to write for him. J.I. Rodale was known for bringing together organic farming pioneers and spreading their knowledge through his publications.

In 1945, Rodale wrote and published the book *Pay Dirt* in which he describes organic farming methods and the dangers of chemical farming. He established the Soil and Health Foundation in 1947 on a 63-acre farm in Emmaus Pennsylvania, later renamed the Rodale Institute. He was a prolific author and publisher of organic farming and health books and magazines including the *Encyclopedia of Organic Gardening* and *How to Grow Vegetables and Fruits* by the organic method. J.I. Rodale left a huge legacy and is considered the father of the American organic movement. It wasn't until after Rachel Carson's book *Silent Spring* launched the environmental movement and the growing counterculture movement took notice of his work that J.I. Rodale found much of a following. But by 1971, his circulation of organic and health magazines had increased significantly. He was featured on the cover of *New York Times*

Magazine in an article published on June 1971 called "Guru of the Organic Food Cult," the day before he died of a heart attack.

Robert Rodale (1930–1990)

Robert Rodale is the son of J.I. and Anna Rodale and a vocal proponent of the organic movement. He was born in New York shortly before they moved to Emmaus, Pennsylvania. He grew up farming with his father and studied English and journalism at the University of Bethlehem, PA. He married his wife Ardath in 1951, and they had five children together. He joined his father at Rodale Press as an editor. During that time, he was influenced by the writings of Charles Darwin, F.H King, and Albert Howard. Robert became president of Rodale Inc. in 1954, and he managed the press for his father who had mostly retired. He sent a member of the Rodale staff to the founding meeting of IFOAM in France in 1972. He took over as CEO of both Rodale Press and Rodale Institute in 1978. Robert expanded the work of the Rodale Institute by purchasing a 333-acre farm in Kutztown, Pennsylvania, in 1971 shortly after his father's death. There he initiated long-term research trials on organic and conventional farming that have given scientific credibility to the organic movement. He helped start the *New Farm* publication geared toward large-scale farmers. He authored a number of books including *The Basic Book of Organic Gardening.* Robert took the small company with an alternative hippie image and made it a well-respected publishing house and research institute. Many people in the organic farming sector credit Robert with bringing the organic sector into the public realm. Robert was killed in a car crash in Russia in 1990 at the age of 60 while on a trip to start a Russian version of the *Organic Gardening* magazine. Robert felt his purpose in life was to expand the organic movement and transform America. His wife, Ardath Rodale took over as CEO and chairwoman after his death. Three of their five children joined the company full-time.

Maria Rodale (1962–)

Maria Rodale is the third generation of Rodales to lead the family's publishing company. She joined the company in 1987, and then in 1998, she became director of strategy. She was editor-in-chief of the *Organic Gardening* magazine and oversaw the organic living division of Rodale Press. She published a number of books including *Organic Manifesto: How Organic Farming Can Heal Our Planet, Feed the World, and Keep Us Safe, Maria Rodale's Organic Gardening,* and *Maria Rodale's Organic Gardening Companion.* She also served on the board of the Rodale Institute. She took over as CEO in 2009 after the death of her mother. Maria has continued her family tradition of advocating for organics, this time from the perspective of the consumer.

Vandana Shiva (1952–)

Vandana Shiva is an activist, writer, and advocate for issues relating to globalization, biodiversity, and ecofeminism. Vandana was raised in Dehradun, a large city at the foothills of the Himalayan mountains in northern India. She developed a strong connection to the forests and ecosystems surrounding her hometown through the work of her parents. Her father was a forest conservator and her mother a farmer.

Shiva graduated from Punjab University in 1972 with a Bachelor of Science in physics. She moved to Canada in 1977 and completed a master's degree in the philosophy of science at the University of Guelph and then a doctorate in the philosophy of physics at the University of Western Ontario in 1979. During her visits home throughout her university years, she volunteered with the Chipko movement, a protest against the deforestation of the forests in the Himalayas led by village women. The Chipko women were the original literal tree huggers, and Vandana's experiences with them influenced her life's path. This is where she first encountered the idea of ecology and biodiversity providing a basis for local economies. She became a proponent of ecofeminism and eventually coauthored the book *Ecofeminism* in 1993. Her engagement with the Chipko movement had a profound impact on Vandana and set her on the path of environmental activism.

After graduating with her PhD, Vandana returned to India and worked at the Indian Institute of Science and the Indian Institute of Management. In 1982, she formed an independent institute, the Research Foundation for Science, Technology and Natural Resource Policy. There she focused her efforts on addressing ecological and social issues in partnership with local communities and social movements. Out of her work with the institute grew a project to save heritage varieties of seeds. This led to the creation of a movement called Navdanya (nine seeds) in 1991. The movement is dedicated to protecting the integrity and diversity of living resources through the promotion of organic farming and fair trade. Navdanya supports an organic research and teaching farm and a collection of community seed banks, and more recently, it has set up a fair-trade organic supply chain in the local region.

Dr. Shiva has made her life's work about resisting the conventional approaches to agriculture. Her work on a global scale has uncovered many of the issues facing organic farmers and has inspired many people to do similar advocacy work in the United States. She advocates on behalf of organic food systems through her teaching and speaking engagements and by authoring books and articles that put a spotlight on issues of biopiracy, intellectual property rights, GMOs that threaten biologically diverse seeds, farmers' rights to save seeds, and organic farming systems. In addition to these issues, she has also been a very vocal opponent of pesticide use, GMOs, and globalization. Shiva believes that human and environmental rights are linked together. She draws a link between the growth of modern agriculture and farmer suicides in India, suggesting that depending on larger corporations for seeds, fertilizers, and pesticides have put farmers in debt. She explains that globalization in the agriculture sector has led to a loss of indigenous knowledge and ownership over seeds that had been bred for generations and subsequently a loss in biodiversity.

In 2015, Vandana Shiva worked with others in the global sustainable agriculture community to found Regeneration International, an organization dedicated to promoting regenerative agriculture practices as a solution to the climate crisis, world hunger, and problems facing the world's social, ecological, and economic systems. This nonprofit organization has provided inspiration and laid the groundwork for many of the organic organizations taking up regenerative agriculture as the newest trend in certification and farming approaches in the United States.

Her answer to these challenges is one of local economies using agroecology principles, especially those of organic agriculture and fair trade. Vandana has led campaigns and lawsuits against larger corporations and government agencies to try and fight against these globalization practices. Vandana Shiva works with governments around the world to develop sustainable agriculture policies and programs and participates with numerous organizations as an invited expert or board member. She has published more than 20 books on the subject and over 300 articles in both scientific and popular journals and magazines.

While Dr. Shiva has received her fair share of negative media and criticisms, she has also been honored with numerous awards and honorary degrees, and she was even called an "environmental hero" by *Time* magazine in 2003. Her broad scope of work has been recognized with several peace prizes, and in 2010, Forbes Magazine named Vandana Shiva one of the Seven Most Powerful Women on the Globe. Much of her work, beliefs, and accomplishments are highlighted in an upcoming documentary called *The Seeds of Vandana Shiva.* By 1993, her work to protect farmers, biodiversity, and the ecological systems was recognized on a global scale, earning her a place on the Global 500 Roll of Honor from the United Nations Environment Programme, an Earth Day International Award, and a Right to Livelihood Award (also known as the

Alternative Nobel Prize for human rights). She has appeared in a number of documentary films including *Dirt! The Movie, The World According to Monsanto, Queen of the Sun,* and many more. The following publications are just a selection of Vandana Shiva's books that most closely address issues related to organic agriculture.

Monocultures of the Mind: Perspectives on Biodiversity and Biotechnology, 1993
Biopiracy: The Plunder of Nature and Knowledge, 1997
Stolen Harvest: The Hijacking of the Global Food Supply, 2000
Tomorrow's Biodiversity, 2000
Patents, Myths and Reality, 2001
Manifestos on the Future of Food and Seed, 2007 (editor)
Monocultures of the Mind: Perspectives on Biodiversity, 2011
Soil Not Oil: Environmental Justice in an Age of Climate Crisis, 2015
Who Really Feeds the World? The Failures of Agribusiness and the Promise of Agroecology, 2016

Rudolf Steiner (1861–1925)

Rudolf Steiner is given credit for being the founder of the biodynamic movement, an approach to agriculture that emphasizes the holistic, ecological, and spiritual elements of farming. He was born in what was at the time part of the Austrian-Hungarian Empire and present-day Croatia. He studied at the Vienna Institute of Technology. He started his career by editing works by Goethe and others. He began writing his own philosophical works in 1886. In 1891, Rudolf Steiner earned a PhD in philosophy at the University of Rostock. Steiner was involved in a wide variety of activities that integrated philosophy, spirituality, and mysticism with science. He wrote and lectured on numerous topics relating to health, education, and the arts with the Theosophical Society. He then founded the Anthroposophical Society in 1913 to support his spiritual-scientific approach to knowledge.

Despite Steiner having no background in agriculture or food production, a group of farmers who had read and listened to his lectures on health, economics, and education asked if he could provide some insights into how they could revitalize their farms in the era of chemical intensification. Steiner responded in 1924 with a series of eight lectures held in Koberwitz, Germany (now Poland). Rather than teach experienced farmers about farming practices, he introduced the concept of pairing holistic farming practices with a spiritual understanding of nature. He emphasized working in harmony with natural cycles and maintaining a view of the farm as a single system. In his lectures and writing, he never used the term "biodynamic" or "organic." Those terms were created by his followers after he died. Steiner explained that his views were not prescriptive and that they required experimentation. The results of the lecture were the organizing of an agricultural group within the Anthroposophical Society. Under the direction of Ernst Stägemann and Dr. Ehrenfried Pfeiffer, this group conducted research and developed agricultural practices and principles that now form the basis of biodynamic farming. Rudolf Steiner died in 1925, not long after giving the lecture series. His teachings inspired a group of farmers to create an organization and a set of standards for farming biodynamically. The Demeter organization, founded in 1928, was the first to create a certification system for a particular method of production. In 1929, the course was translated into English by George Kaufmann, a Cambridge University graduate who spent a significant amount of time with Steiner before his death. The series was published as a book, *Agriculture: Spiritual Foundations for the Renewal of Agriculture*.

Organizations

California Certified Organic Farmers

California Certified Organic Farmers (CCOF) was formed in 1973 with 54 farmer members who wanted to create an organic certification. CCOF comprises three different entities that are each structured in a different way. CCOF Inc. forms the initial membership organization, now with 14 chapters and an elected board of directors. In 1975, CCOF launched their first chapter, the Central Coast Chapter. The chapters allowed local farmers to connect with one another in person several times a year and to address issues related to their region. This part of the organization provides the main structure of the CCOF. The CCOF Certification Services, LLC, was created to run the organic certification programs that include both domestic and international certification options. Finally, they operate the CCOF Foundation that manages a number of funds for education and farmer support. The OFRF was formed by the CCOF to raise funds and support research that focuses on growing the organic sector.

The CCOF is now a large and influential organization, but when they started in 1973, they were just one membership organization, made up of farmers who developed a certification process based on 13 rules. The first organic certification in the country was run by Rodale, but in 1973, they announced that they would no longer be running certification operations in California. At that point, those who had been certified by Rodale decided it was time to create their own local certification program. To start, members inspected each other's farms, while at the same time learning and sharing knowledge. Until 1985, they were entirely run by volunteer farmers, but they had managed to certify nearly 150 farms. By 1987, CCOF had published their first certification handbook and a farm inspection manual. And they began the process of training farm inspectors. After the 1989 Alar scare, CCOF expanded to include more than 800 certified organic growers across the state. As they grew, they decided they could no longer manage as a volunteer-run organization. Their first staff members, Mark Lipson and Bob Scowcroft, have worked for many years in the organic sector, in CCOF, OFRF, and with other organizations to support the growing industry.

CCOF advocated for a California organic food law, which was finally passed in 1990 the same year as the OFPA. The passage of the law again led to the increase in the number of operations that CCOF certified. CCOF and Oregon Tilth were struggling to keep up with the long and complicated list of possible inputs that could be approved for use on organic farms. So, they supported the creation of a third-party organization dedicated to reviewing materials for use in organic farming. That organization is now known as OMRI and used widely by most certifiers and farmers.

They also expanded their reach by developing a certification for processed food and even opened a membership chapter specifically for processors as the processed organic food industry was growing rapidly. By 2006, they were the largest certifier in the United States, certifying over 11 percent of all organic businesses. The organization has been a leader in developing online tools to streamline the certification process for farmers. As the organic industry has evolved, so has the certification agency. They now certify everything from growers to ingredient suppliers, handlers, packers, processors, warehouses, private labels, retailers, and restaurants. As California is a huge participant in the organic import and export market, the CCOF has developed a full range of international certification programs for exporting goods around the world, first by becoming accredited by IFOAM in 1998 and eventually by creating their own equivalency programs. They have even expanded into Mexico, complete with their own chapter of the CCOF. They also offer a transition program to support farms as they transition into fully certified operations.

While CCOF maintained their primary business of certification, they began taking on more and more advocacy work. This led to the current structure with separate entities for certification and advocacy work. Over the years, the organization has advocated for stringent organic rules at both local and regional levels and advocated against GMO crops. They follow the work of the NOSB and regularly comment on how the rules will impact their members. One of their big successes was the passage of the California Food Products Act in 1999, which served as the foundation for the NOP standards. In 2008, they created a dedicated staff position for policy and advocacy work. CCOF also has a dedicated action fund to support their advocacy work.

The CCOF Foundation is supported through a kickback program associated with certification. Each certification fee includes 2 percent that goes directly to the foundation to support education, training, and advocacy work. With funds from the foundation, CCOF developed an organic training institute to host hands-on training, workshops, webinars, and seminars. The foundation also manages a Future Farmer Grant Fund to support organic education for students from kindergarten through university or vocational training. They run a hardship assistance program to support organic farmers who experience economic hardship.

The foundation also runs a consumer education and awareness program on the benefits of buying organic food. And to support the growth of organic agriculture, they recently analyzed over 300 scientific studies and compiled the results in a report titled "Roadmap to an Organic California: Benefits Report." A second report will be released in 2020 and will outline a series of policy recommendations to increase organic acreage in California from 4 percent to 10 percent by 2030.

CCOF regularly collaborates with other organizations. They are a founding member of the California Climate and Agriculture Network, and they work closely with the Organic Trade Association and the National Sustainable Agriculture Coalition.

International Federation of Organic Agriculture Movements

The International Federation of Organic Agriculture Movements (IFOAM) is an umbrella organization that brings together a diverse range of stakeholders to support organic progress around the world. The organization creates awareness, advocates for supportive policies, and facilitates the creation of organic value chains. The IFOAM's mission and goal is to support worldwide adoption of systems based on their principles of organic agriculture. The organization's principles of organic agriculture are based on the principle of health, ecology, fairness, and care. This set of principles are both drawn from organic agriculture and suggest what organic agriculture can offer the world. They suggest that the health of people cannot be separated from the health of soils, crops and animals and that organic agriculture should be based on living cycles, built on fair relationships, and managed responsibly to protect current and future generations.

In 1972, Roland Chevriot of Nature et Progrès, an organic organization located in France, invited many organic organizations from around the world to meet. Four other organizations joined him in a meeting in Versailles, France, to discuss a coalition dedicated to advancement of organic agriculture worldwide. In attendance at the first meeting was Lady Eve Balfour of the UK Soil Association, Kjell Arman from the Swedish Biodynamic Association, Pauline Raphaely from the Soil Association of South Africa, and Jerome Goldstein from the Rodale Institute. The organization was volunteer-run until the mid-1980s when they hired their first staff member. In the beginning, the focus of the organization was on science and farmers. As membership grew, so did the issues that concerned the organization. When the organic sector grew to include a large business contingent, these organic businesses wanted to join IFOAM as well. Many in the organization were worried that this would erode the core issues and values of the organization. Eventually, the organization created a distinction between organizations that were predominately organic versus those that were not, rather than business versus nonprofit. They also created an associate category that did not include voting rights. Thus, the organization was able to draw in more membership fees without risking a takeover from outside interests. Currently, the organization has over 800 members (or affiliates as they are called) in more than 120 countries, including 67 located in North America. These affiliates elect a World Board at the General Assembly, form committees, and join regional alliances. The regional alliances are often formalized as Regional Bodies of IFOAM working on specific regional or sector priorities.

In 1980, IFOAM organized a committee to draft the IFOAM Basic Standards. These standards have influenced the development of standards around the world, including the EU regulations and those developed by Codex Alimentarius (part of the UN FAO and the World Health Organization). The standards led to the creation of the IFOAM Accreditation Programme in 1992. The initiative was created as a separate program called the IFOAM Organic Guarantee System. Through this program, IFOAM initiated a harmonization process to facilitate international trade of certified organic products. The program evaluates certification programs around the world and then give accreditation to the ones that meet their minimum requirements. They also maintain a list of the certification programs that have adopted the IFOAM standards. As demand for organic products grew, new issues arose, including the impacts of certification requirements for small-holder farmers in developing countries. IFOAM joined forces with UNCTAD and FAO to create COROS (Common Objectives and Requirements of Organic Standards), a tool for assessing equivalency of standards.

In 1992, IFOAM decided it was time to develop a relationship with the FAO, and in 1997, it was granted observer and liaison status with the United Nations and the relevant sub-programs including FAO, UN Conference on Trade and Development (UNCTAD), International Labour Organization, the UN Environmental Programme, Economic and Social Council(ECOSOC), Organization for Economic Cooperation and Development(OECD), Codex Alimentarius, and the International Organization for Standardization (ISO).

The lobby efforts that IFOAM initiated with the UN and the FAO especially have had significant impact. The FAO now lists organic agriculture among its top five priorities.

In 2017, IFOAM launched Organic 3.0 to bring organic out of a niche market into a broader framework for tackling complex sustainability issues worldwide. Organic 1.0 was the origin of the organic movement about 100 years ago. Organic 2.0 saw the organic sector grow and the subsequent codification and standards creation in over 87 countries. Now, Organic 3.0 will address three areas where organic 2.0 has failed. The first is to support the small organic producers who are excluded from organic certification. The second is to link organic agriculture to other sustainability initiatives such as agroecology, fair trade, and other related issues. The third is to address the economic pressures that push organic producers to drop some of the core organic practices such as diversification in favor of increasing economies of scale.

In 2018, IFOAM had three main focus areas: facilitating development of the organic sector through training for transition farmers and future leaders and encouraging innovative solutions through think-tank work; raising awareness by organizing consumer campaigns, hosting events, and publishing resources in multiple languages; and advocating for policies that protect organic and address food security, climate change, and biodiversity. They partner with organizations around the world to run programs to advance organic value chains especially in developing countries. One of the main activities of IFOAM is bringing people together for their biennial general assemblies and the scientific conferences that they created along with them. Over time, the organization started organizing and participating in many additional conferences on a wide range of topics related to organic. IFOAM publishes a magazine, *Ecology and Farming*, a newsletter *IFOAM—In Action*, conference proceedings, reports, and position papers.

Maine Organic Farmers and Gardeners Association

The Maine Organic Farmers and Gardeners Association (MOFGA) began in 1970 when farmers who were curious about organic farming started gathering to learn from one another. Similar to many farming organizations, it began as an informal way for farmers to share information and learn what they could in a time when there was little available. After several meetings together, the organic growers decided to create an official organization. They got early support from an agricultural extension agent with the University of Maine, who gave them space and support to produce a newsletter.

By 1972, a number of farmers came together to organize an organic certification program based on the Rodale guidelines. They began by certifying 27 farms in their first year. In 2002, they separated the certification agency from their main organization in order to maintain an arm's length from their advocacy work. It is now called MOFGA certification services and certifies farms in Maine, New Hampshire, Massachusetts, and Vermont.

By 1995, MOFGA had become the largest state-level organic farming organization in the country. By 2010, they had a membership of over 6000, with about 1500 coming from outside of Maine. While similar in many ways to other organic farming organizations, they have a few very unique offerings that make them one of the most popular organizations.

In 1977, MOFGA organized the Common Ground Country Fair, held in Litchfield, ME. It was the first country fair dedicated to organic and sustainable rural life. The fair drew a crowd of 10,000 people that first year and raised enough funds to help the organization grow. They have continued to host the fair each year with more than 60,000 people coming from all over North America to attend. This unique effort has helped them fund and grow other distinct programs.

MOFGA was able to expand its educational programs with the purchase of over 300 acres in Unity, ME, where they created a year-round training and research center in 1996. The

Common Ground Education Center is designed to host everything from the annual country fair to conferences and a farmer-in-residence program. The grounds are used as demonstration and teaching plots with farm fields, gardens, trees, and high-tunnel greenhouses. They have developed an experimental orchard at the Unity farm, where they grow 500 different varieties of apple and pear trees. Their library has nearly 2000 books on sustainable and organic agriculture and numerous magazines and other periodicals. Another special feature is the large commercial kitchen that they allow farmers to use to process food into value-added products to sell.

In 1986, MOFGA became the first organic farm organization to hire their own extension agent, a staff position dedicated to providing technical assistance to organic farmers. They now have one of the most comprehensive agricultural services available for organic farmers, complete with training and support on every aspect of organic production, business, and marketing. They have expanded to maintain two full-time staff dedicated to providing technical assistance.

In 1990, they hosted their first farmer-to-farmer conference with 82 participants. The conference is now held annually, along with numerous other workshops, seminars, and training events. Because Maine is a heavily forested state, they recognized that many farmers are also managing forestland, and they began developing programs for low-impact forestry. Besides technical support, MOFGA offers financial assistance in the form of grants to farmers who are interested in receiving consulting or technical assistance.

In 1975, the organization launched a formal apprenticeship program despite having little funding for programming. By 1999, they had hosted over 500 apprenticeships and started a journeyperson initiative as a bridging program to owning a farm. They have since then developed a full suite of new farmer training sessions designed to take beginners through four stages of training including apprenticeship, journeyperson, farmer-in-residence, and farm beginnings to support farmers through the challenging early years. In addition, they host a Farm Resilience Program designed to support mid-career farmers in scaling up their business or reaching new markets. And they support farmers with a farm equipment sharing program.

MOFGA decided to expand their offerings beyond farmers with more educational programs for home gardeners as a way to expand their reach beyond the farming community. They also began working with farmers' markets and community gardens to make local organic food more available for low-income families. They publish a number of food guides to help consumers find local organic food in their communities. And they supported the creation of a school garden network throughout the state to bring organic farming practices to a wider audience.

MOFGA publishes a regular newsletter, a separate newsletter specifically for their certified organic members, wholesale price reports, a weekly pest report, and occasional technical reports. They also host a weekly radio show called the *Common Ground Radio Show* in conjunction with the local community radio station. Furthermore, they became involved in policy work by supporting a no-spray and pesticide drift campaign and by advocating for organic food labeling. MOFGA continued their advocacy work with successes such as making Maine the only state with a ban on rBGH (for a time anyways, as the hormone was later approved). They provide policy statements and advocate for both federal and state policies that support their organic members. They provide policy position statements to encourage and support their members in taking their own actions and host events to support community action on important issues. They participate in coalitions including the National Organic Coalition.

Northeast Organic Farming Association

The Northeast Organic Farming Association (NOFA) is an affiliate association with over 5000 members across seven state chapters: Connecticut, Massachusetts, New Hampshire, New Jersey, New York, Rhode Island, and Vermont. NOFA supports the work of each individual chapter to advocate and educate on organic agriculture. The umbrella organization publishes a quarterly newsletter called *The Natural Farmer*, organizes an annual summer conference, and supports programs that stretch across multiple chapters. Two such programs include the Soil Carbon Restoration project, which is focused on providing farmer education on soil health and climate mitigation strategies, and the Beginning Farmer program dedicated to supporting new farmers with scholarships, targeted education, mentoring, and apprenticeship.

NOFA was founded in June 1971 by Samuel Kaymen in Westminster, Vermont. Samuel Kaymen, who went on to create Stonyfield Farm Yogurt, gathered together a group of like-minded individuals to learn about organic farming practices by forming study groups, sharing information with one another, and finding new markets for their products. These informal farmer meetings evolved into regular workshops and a full suite of farmer training and support. One of their first major events was a conference held in Wilton, New Hampshire, in the summer of 1975. There were 350 attendees who came to hear Wendell Berry speak at High Mowing Farm. This conference became an annual event that farmers traveled to from all over the world. In 2019, they held their 45th annual summer conference in Amherst, MA. Each state chapter also hosts their own smaller winter conference each year.

Another of their early activities was distributing food grown on members' farms to food coops, natural food stores, and other consumers by organizing truck routes to New York City. It didn't take long for them to decide that it was much more complicated that it seemed, and this quickly changed to a focus on local food initiatives such as farmers' markets and wholesale grower cooperatives.

In 1979, NOFA launched one of the earliest organic certification programs in the country. The organic standards developed by NOFA served as a model for many others in the country. Grace Gershuny who developed the standards for NOFA went on to help write the rules for the NOP. The organization was also one of the founding members of the OTA and helped develop that organization's guidelines for the organic industry. It currently operates the Northeast Interstate Organic Certification Committee that supports chapters that offer USDA certification, including NOFA-New York, NOFA-Massachusetts, NOFA-Vermont.

NOFA also collaborates with other organizations on issues of advocacy and social justice and provides policy guidance to the chapter memberships. The organization attends NOSB meetings and comments on issues that impact their membership. NOFA has played a strong role in the creation of many supporting organic organizations. They are founding members of the National Organic Coalition, Domestic Fair Trade Association, National Sustainable Agriculture Coalition, and the Agricultural Justice Project. They are also represented in larger organizations such as the North East Sustainable Agriculture Working Group and IFOAM.

In keeping with their mission to provide education and training to organic farmers, NOFA has published guidebooks on various aspects of organic farming practices: *Organic Soil-Fertility and Weed Management*; *Crop Rotation and Cover Cropping*; *Compost, Vermicompost and Compost Tea*; *Growing Healthy Vegetable Crops*; *Organic Dairy Production*; *Organic Seed Production and Saving; Whole Farm Planning*; and *Humane and Healthy Poultry Production.*

Each of the seven chapters has its own programs and focus that depend on the needs and volunteer resources available in that state.

NOFA-Connecticut puts a large emphasis on connecting consumers with local food providers and even organic landscapers. They publish an annual farm and food guide and an interactive map to help consumers connect with farmers, farmers' markets, and other sources of local organic food. They also run a program called Organic Land Care that encourages organic lawns, gardens, and landscapes and includes an accreditation course for land care professionals. Support for farmers includes on-farm workshops, mentorship, and consulting. They also host a Farmer's Pledge in which farmers can commit to a set of principles.

NOFA-Massachusetts has a robust farmer training program, a bulk ordering program for farmers and gardeners including a significant program on soil health, a food guide to connect consumers with farm products in the state, a food access program, and advocacy work at the state level including a local raw milk campaign. They also participate in the Organic Land Care program and the Beginning Farm Program.

NOFA-New Hampshire offers farmer training including apprenticeships and special training on profitability. They participate in the Organic Land Care program and provide training for backyard gardeners. They also run a bulk purchasing program for members. In addition, they have several programs dedicated to supporting low-income people with local organic food directly or through gleaning programs to support local food pantries and soup kitchens.

NOFA-New Jersey offers training and support to organic farmers and gardeners by hosting local events and through their quarterly newsletter. They recently participated in a farm to school program.

NOFA-New York offers farmer training on production, marketing, value-added processing, and technical assistance. They also run an organic certification program and include a 100 percent grass-fed certification. They work on advocacy issues in their state and promote organic products to consumers including a farm and food guide.

NOFA-Rhode Island offers education and training and marketing support and participates in the Land Care Program and the bulk ordering program. They have initiated an equipment sharing program for farmers in partnership with Johnny's Selected Seeds.

NOFA-Vermont has developed a range of programs to service both farmers and consumers. They have programs to address community access to food, and they have established farm to school programs and farm to institution programs to infuse schools, colleges, hospitals, and other public institutions with local organic food. They offer a range of support to local farmers' markets and even have a mobile pizza oven to promote local organic food, in addition to their farm and food directories. They offer a full range of farmer training and support services for farmers at all stages of their careers on everything from production to farm business and marketing strategies and apprenticeships. They have even developed two loan programs for their membership. And they run an organic certification program.

National Organic Coalition/Rural Advancement Foundation

The National Organic Coalition is a suborganization of the Rural Advancement Foundation International (RAFI-USA). The RAFI-USA has existed in some form since the early 1900s. Originally formed to advocate for family farms and rural communities, its early efforts were focused on sharecropping and tenant farm issues. In the late 1970s, the organization began to focus on the loss of plant genetic diversity and expanded to include an international focus. Over time, the organization formed a program on sustainable agriculture. Currently, the program on sustainability, called Just Foods, covers a range of issues including genetic diversity, organic plant breeding, maintaining organic integrity, food justice, and pollinator protection.

The bulk of their support for the organic sector has come through the Organic Integrity program, which first helped form the Sustainable Agriculture, Research, and Education program, funded by the USDA and tasked with granting funds for research and education. RAFI-USA also played an advocacy role in the creation of the OFPA. The former program director at RAFI-USA, Michael Sligh, served as the first chair of the NOSB. RAFI-USA's Seeds and Breeds program advocates for policies that protect public plant breeding, collaborate and build community, and support seed research programs. In their Breeding for Organic Production Systems program, they connect farmers and breeders to field test crop varieties and host an annual meeting to bring researchers and farmers together to discuss the needs of the sector.

Another major effort of the Just Foods program was the co-creation of the Agricultural Justice Project. They partnered with several farmworker and organic farming organizations that were disappointed that the NOP did not address any farmworker justice issues, and they decided to create a Food Justice Certified label that maintains standards for fair and respectable treatment of farmworkers.

The organization then determined there was a need for a farmer-focused coalition of farmers, farm organizations, and other nonprofits dedicated to organic integrity. In 2003, they formed the National Organic Coalition (NOC) to lobby for organic farmers at the federal level. The NOC is run independently of RAFI-USA, but it's supported financially in part by donation and in part by RAFI-USA. The organization employs a dedicated policy director and is membership based. Membership is by invitation and has three stakeholder categories: farmers and farm organizations; nonprofit, consumer, environmental, and animal welfare organizations; and businesses dedicated to organic integrity. Policy decisions are made by consensus in the membership base.

The NOC currently has 14 member organizations and a network of approximately 30 entities that are considered network affiliates who are invited to participate in events and receive updates on policy actions. The coalition also draws on a number of advisors to help draft policy and plan advocacy efforts. While most of their efforts are put into lobbying Congress and providing testimony to the NOSB, they do occasionally join other organizations in lawsuits against the USDA when other actions have failed on a significant issue. The most recent example is when they joined with others to sue the Trump Administration for withdrawing the Organic Livestock and Poultry Practices rule.

In 2019, the NOC had eight priority policy areas that include appropriations (convincing Congress to spend more on organic agriculture); advocating for the NOP every time the Farm Bill is up for renewal; keeping up with the work of the NOSB and providing testimony on important issues; advocating for organic rights around GMO issues; lobbying for access to seeds and breeds for organic production, funding, and support for organic research; exploring challenges to expanding U.S. organic production; and lobbying for effective enforcement of the standards to ensure organic integrity. The NOC claimed success in many of their policy areas in the 2018 Farm Bill, including an increase in NOP funding and staffing; progress on the organic pasture rule; an increase in organic research funding through the Organic Agriculture Research and Extension Initiative; and funding for organic certification cost-share programs, enforcement, and organic data collection.

The NOC has taken on much of the advocacy work on organic issues, but the RAFI-USA still plays an active role in coordinating national organic issues. In 2008, they convened a series of dialog meetings with over 300 stakeholders across the country to hear what the sector needed to move forward. The results were drafted into a document titled "Towards a National Organic Action Plan." In 2009, they held a national summit with over 85 participants to review the results and draft a *National Organic Action Plan* outlining a clear vision and plan for the future of organic food and agriculture. This was used to establish a set of goals and benchmarks to track progress. The ultimate goal of the project is that organic should be the foundation for food and agriculture production systems in the United States. The plan was modeled on the organic action plans found in many EU countries, but with one main difference: instead of a government-originated plan, this one originates with the grassroots organic sector.

Oregon Tilth

Oregon Tilth has its roots in a group of farmers who met in 1974 after a symposium for small farming that hosted Wendell Berry as the keynote speaker. After the symposium, Wendell Berry wrote a letter to some of the people he met there, suggesting that change in agriculture would only come from a grassroots effort and that another symposium would help along those efforts. The conference was organized for the following year, and from that effort, a newsletter was started. In 1977, several individuals decided to form a membership organization to support organic farmers and launched Tilth Association to represent organic farmers in the Pacific Northwest. An overlapping group of commercial organic farmers also formed the Tilth Producers' Cooperative. A number of regional chapters of the Tilth Association were formed up and down the coast. Together they published a number of books, hosted conferences, and expanded membership. But by 1984, the volunteer-run organization had run out of resources, and they decided to split into individual groups.

For many years, there was a loose collection of informal tilth organizations in the Pacific Northwest. In 2016, Seattle Tilth, Tilth Producers, and Cascade Coalition became a more formal organization of about six to seven active chapters collectively called the Tilth Alliance, including the Tilth Producers chapter that was formerly the Producers' Cooperative. Over the years, the Tilth Alliance has created a unique array of programs to serve both rural and urban communities. They manage community gardens and educational farms, offer training and support to urban farmers, host a harvest fair and a major conference for commercial organic growers, and offer a range of programming for kids. They helped develop the Washington State Department of Agriculture's first state-managed organic certification program.

Oregon Tilth was created in 1984 when the initial organization was disbanded. They initiated the launch of one of the earliest, and now most well known, organic certification services. By 2007, Oregon Tilth was certifying approximately 600 farms and 600 food processors, and it has since expanded to certify farms across North and South America. They were also the first to develop a certification process for restaurants in partnership with Restaurant Nora in Washington, DC. The Oregon Tilth standards developed in the early 1980s have become the basis for state certifications in Washington, Oregon, Texas, Idaho, Colorado, Hawaii, and Japan. They were also used in the creation of the NOP standards. In 1992, Oregon Tilth and CCOF worked together to create a joint materials evaluation program that become a separate entity in 1997: The Organic Materials Review Institute.

While certification services are the core of their programming, Oregon Tilth remains a membership organization dedicated to supporting organic farmers and others in the organic sector with over 1800 members from all aspects of the supply chain. They offer education and support for farmers transitioning to organic and those currently farming organically. The organization publishes a quarterly magazine called *In Good Tilth* with a circulation of 9000. They co-host Organicology, a national conference for farmers, processors, and retailers, along with the Organically Grown Company, the Organic Seed Alliance, and the Sustainable Food Trade Association.

Oregon Tilth has decided to focus a significant portion of their support efforts on transitioning farmers who are currently operating in the conventional sector to an organic system. They know that the three-year transition period can be difficult and includes a steep learning curve. To address that, they have developed a number of programs geared toward farmers who are ready to transition including webinars, workshops, and seminars. They also encourage peer support through networking events that bring together all stages of the supply chain. In 2016, they decided to survey farmers who have transitioned to organic practices and undertaken certification to identify challenges and new areas for support. The results of the survey, along with financial support from Oregon Tilth, are informing the development of a new Organic Extension Program at Oregon State University (OSU).

Oregon Tilth has also partnered with the USDA Natural Resources Conservation Service (NRCS)to educate the NRCS staff on ways to use organic farming practices to meet conservation goals. They have created a series of trainings for people who work with farmers to create conservation plans and have published a National Organic Farming Handbook on how to implement conservation practices on organic farms.

The organization has expanded their work beyond the production side of things, and it now offers a series of business management skills training. In addition to their own technical assistance to farmers, they have partnered with OSU to create a course specifically designed for whole farm management strategies as part of their organic extension program. Through their partnership with OSU, they also created an organic fertilizer and cover crop calculator to help farmers assess their on-farm needs.

In 2016, Oregon Tilth decided it was time to create a social investment program to partner with organizations that are committed to equity in the organic sector. They provide funds to support programs for those who have been historically underserved in the agricultural community. Continuing with their roots in advocacy work, Oregon Tilth advocates at the federal level for increased spending on organic research, conservation efforts, and organic integrity in the NOP.

The Organic Center

The Organic Center was founded in 2002 by the OTA to function as an independent nonprofit, based in Washington, DC. Their mission is to conduct evidence-based scientific studies on the environmental and health impacts of organic food and farming to lend credibility to the organic sector. It was first led by

Dr. Charles Benbrook and more recently by Dr. Jessica Shade. The organization further aims to educate and communicate the results of the research to the general population. They engage in partnerships with universities, research institutions, government agencies, and other nongovernmental organizations working in the organic sector. They provide a broad range of resources targeted at consumers, farmers, policymakers, and other researchers on topics related to the science of organic farming.

The organization is headed by a director of science programs and a programs manager. Their work is supported by a board of trustees who direct the organization and a large group of scientific counselors who advise the Organic Center on a wide range of research topics, many of them conducting organic research themselves. The Organic Center has three main programs: Research Projects, Annual Benefits Dinner, and the Organic Confluences Summit. In addition, their website is designed for outreach and education with informational publications, videos, recipes, infographics, a blog, and a newsletter. As part of their effort to share evidence-based information about organic farming and dispel myths perpetuated by media, the Organic Center regularly publishes detailed responses to articles published in mainstream media and posts them on their blog and shares them through social media. The organization also regularly compiles scientific literature reviews and summaries to support the lobby efforts of the OTA. It represents the organic sector at national scientific gatherings and hosts forums with farmers and researchers to ensure research is meeting the needs of the sector both in practical applications for farmers and in gathering evidence of organic farming's success.

The main focus is on the research projects that are supported by the Organic Center, primarily in collaboration with university researchers. To date, the organization has published, collaborated in, or supported 22 projects in some capacity. Most research projects have focused on improving organic production methods to maximize various environmental impacts such as pest reduction, soil building, and reducing pesticide exposure. They also work on studies that provide evidence that organic farming is beneficial for a variety of environmental issues from biodiversity and pollinators to carbon sequestration, and reduction of contaminants ranging from nitrogen pollution to arsenic. They also have a range of research programs related to human health due to exposure to pesticides, antibiotics, and other contaminants in food. They have an entire study and related educational website devoted to the benefits of organic dairy products to human health.

Every month since December of 2013, the organization has sent out a monthly newsletter that highlights the key research reports and includes a round-up of research articles or studies relevant to the organic sector. An archive of these newsletters and a few special reports since their inception to the end of 2016 are available online. They have also authored or coauthored nearly 40 reports on organic research and made them available on their website. These range from reports on reducing pesticide exposure to a recently published report on the benefits of organic dairy.

Every year, the Organic Center hosts a fundraising dinner that serves an all-organic meal and includes a keynote address. The dinner draws organic activists, business leaders, and advocates each year. It is held at the annual Natural Products Expo West show and billed as a premier networking event. In recent years, the fundraiser has attracted more than 500 attendees and raised over $450,000 for the organization.

Beginning in 2016, the Organic Center started hosting an annual Organic Confluences Summit at the Natural Products Expo East in partnership with a number of other organizations to address issues at the intersection of research and policy. Each year, the one-day event has a theme that brings together researchers, farmers, policy makers, and other stakeholders for presentations, roundtable talks, and discussion sessions. The focus of their first theme was turning environmental evidence into policy practice, and their most recent theme was on mitigation and adaptation to climate change.

Over the years, the Organic Center has developed a reputation for providing research support and credible information to those seeking answers to some of the most common questions about the benefits of organic farming and issues that need further exploration. Organizations, policymakers, and others regularly request scientific input, research reports, and advice.

Organic Consumers Association

The Organic Consumers Association (OCA) is a nonprofit organization that was founded in 1998 during a large protest against provisions in the initial NOP standards. It started as an organization called Pure Food Campaign, run by Jeremy Rifkin, to lobby the government to require labeling of foods that contain GMOs. It was taken over by Ronnie Cummins and Rose Welch, based in Minnesota, who were active in the organic community. They built the initial membership during the commenting period for the first draft of the organic standards by contacting those who had commented, asking them to become involved. They also worked hard to get the message out to consumers through natural food stores, food coops, and online. They were largely responsible for the unprecedented consumer feedback the USDA received in opposition to the allowance of irradiation, GMOs, and sewage sludge in the NOP standards.

The organization is a registered nonprofit, and their membership base is made up of over 850,000 individuals and businesses that have supported their work or have signed up for their newsletter. The governing structure of the OCA comprises a board of directors with voting rights and a policy board made up of 15 members representing a broad range of issues from around the world. The board of directors is a combination of policy board members and staff. The policy board guides the efforts of the OCA and the work of the staff that oversees numerous programs and campaigns. The OCA's main efforts focus on supporting organic regenerative agriculture, upholding strict USDA organic standards, as well as alternative certification standards, fair trade initiatives, the end of all uses of GMOs and pesticides and industrial agriculture, transparency in the food industry, and universal health care, including support for natural health and wellness.

The organization runs numerous grassroots lobbying efforts, consumer outreach, and campaigns for a wide range of topics related to organic food, health and food safety, and environmental sustainability. They have a website full of articles and information on a wide range of topics, complete with actions that consumers can take to support various issues, including petitions to sign and boycotts of companies that do not adhere to their ideals for organic standards. They send out a regular newsletter, and publish buying guides and blog posts. They work with retailers to hand out print materials and educate consumers on the benefits of buying organic food.

The organization is often criticized for using inflammatory language to make their case, and they have sometimes used dubious research to support their argument. That said, they have generally advocated on behalf of the organic sector using legitimate arguments. They have partnered with other organizations for many lobbying efforts and occasionally for litigation. The OCA also backs a number of organizations including the U.S. Right To Know, an organization dedicated to investigative journalism in the food industry, and the Citizens Regeneration Lobby, an organization dedicated to lobbying for regenerative agriculture. Together these organizations have exposed a number of underhanded dealings of large biotech companies hiring university researchers to publish ghostwritten material in support of GMOs and in opposition to the organic sector.

One of their oldest running campaigns is Millions Against Monsanto, designed to push back against biotech companies that advocate for GMO seeds. They worked on policy campaigns in California, Washington, Colorado, Oregon, and Vermont and at the federal level to support GMO labeling laws. They supported a boycott of companies and brands that were against GMO labeling and helped consumers identify those who supported a labeling law. Through this campaign, they also test products for pesticides and post the results publicly. On some occasions, they have sued companies, including Monsanto for labeling products as natural, when they are contaminated with pesticides.

More recently, the OCA has filed lawsuits against a number of companies including Ben & Jerry's and Tyson Foods for deceptive marketing practices as part of their The Myth of Natural Foods campaign and Dump Dirty Dairy. The OCA was active in leading a boycott against milk products that were sourced from dairy farms that did not comply with the requirement to provide cows with full access to pasture under the NOP regulations.

Since the beginning of the NOP, they have advocated for strict regulations and regularly lobby for particular rules to be passed such as prohibiting carrageenan and tetracycline. In 2004, the OCA joined a blueberry farmer named Arthur Harvey in a lawsuit against the USDA for allowing synthetic ingredients under the organic standards. The lawsuit caused a large internal fight in the organic industry and has had long-lasting effects. They have a number of other campaigns that encourage consumers to choose organic to save the bees and butterflies, protect against climate change, avoid factory farmed animals and industrial food, and support the integrity of the organic standards, among many others. While the OCA has many supporters in the organic sector and among consumers, they are not without controversy as they often draw on weak theories or pseudoscience to support their arguments, which reduces their credibility.

Organic Farming Research Foundation

The Organic Farming Research Foundation (OFRF) grew out of the CCOF, one of the earliest organic certifiers in the country. It was formed in 1990, just after the passage of the OFPA.

CCOF had grown significantly, but it wanted to do more than just certify farmers. Rather than expand their mission, they decided to form a research foundation that would be eligible for a broader range of funding and scope of programs. At first, OFRF was primarily a source of funding for the education and outreach that were undertaken by the CCOF, but within two years, the board of directors decided to conduct a survey of organic farmers to determine what kind of research they should begin doing. They sent out a survey to as many organic farmers as they could find, and along with the survey, they asked for a small donation. That initial ask raised enough funds through individual donors to gain access to foundation grants that allowed them to begin staffing the organization. Bob Scowcroft, the Executive Director of the CCOF, had been trying to serve both organizations for the first few years. He finally resigned from the CCOF and took over the same position for the OFRF. The OFRF is overseen by a board of directors and a secondary advisory board comprising farmers, researchers, and academics. The OFRF has four priority areas where they focus their efforts to support organic farming: research, community, policy, and education.

Research is the primary focus for the foundation, and it started with their first survey on as many organic farmers as they could contact in 1993. That first survey indicated that weeds were a top research priority for organic farmers. The OFRF has continued to conduct national organic farmer surveys to learn about the needs of organic farmers, and they use the results to guide their funding priorities. Over the past 30 years, the OFRF has awarded over $3 million to over 300 research projects on everything from plant breeding to weed management and water use across multiple countries. All the research results are available for free and in a searchable database online.

And unlike most other grants, the OFRF provided research dollars to farmers, not just researchers. Many organic farmers have come to rely on the technical reports that have been generated through OFRF research to improve their production methods. The OFRF is well known for a number of resources that they published in the early days of the organic sector. In 1997, they published *Searching for the "O" Word: Analyzing the USDA Current Research Information Systems for Pertinence to Organic Farming,* which reviewed more than 30,000 projects funded by the USDA and found that only 34 were strongly relevant to organic agriculture. They followed up that report with an ongoing project to track research in land-grant institutions. The second report in 2001 found that 39 states had some organic research, and in 2003, there was only a small increase to 44 states. These reports were catalysts for many of the successes the OFRF has achieved in elevating organic agriculture research and funding.

The education focus is represented through most of their research initiatives by requiring each project funded by OFRF to include an outreach element that brings together both the farming community and other researchers. These are often in the form of farm tours and field days, presentations at conferences or forums, online webinars, and written publications. They maintain a monthly newsletter to connect farmers with the wide variety of resources available. One of their more recent initiatives is an online course titled Organic Soil Health Management, offered for free in partnership with the University of California Sustainable Agriculture Research and Education Program and California Polytechnic State University.

Another area of focus is building a community of support for organic farmers. In 1993, they hosted the first Organic Leadership Conference and continued to host one every other year for a decade to encourage national collaboration and learning. The OFRF wanted to focus their efforts on farmer-led research rather than just university researchers working on test plots and in laboratories. They used several initiatives including one called the Scientific Congress of Organic Agriculture Researchers. It started with two national meetings to bring farmers and scientists together. The initiative, now called the Organic Agriculture Research Forum, continues to facilitate farmer and researcher collaborations with an annual meeting, often in partnership with other organizations. To reach a broader network of farmers and researchers, OFRF often collaborates on research initiatives with other organizations. In 2019, the OFRF partnered with the Organic Seed Alliance and received a National Institute of Food and Agriculture grant through the OREI for their program: A National Agenda for Organic and Transitioning Research.

The OFRF has developed a policy initiative with goals to increase government support for organic agriculture, to ensure land-grant universities have incorporated organic research and education programs, and to support policies that support organic farmers' economic viability and rights. They regularly produce policy statements on proposed legislation that farmers can use when drafting their own comments to proposed rules or when speaking with elected officials. Some of their major lobbying successes include the creation of the OREI in the 2002 Farm Bill and a significant increase in research dollars available for the organic farming community.

Organic Materials Review Institute

The Organic Materials Review Institute (OMRI) is a nonprofit organization based out of Eugene, Oregon, and founded in 1996. OMRI was created to provide independent evaluation of input materials intended for use in organic agriculture. Inputs can include things such as fertilizers, pest control, processing additives, livestock feed and care, cleaning supplies, or any product intended to be used in organic production. They do not review machines or any products that are not relevant under the NOP. It is run by a board of directors with a full staff. An advisory board along with review panels provides scientific oversight. OMRI launched a program in Canada in 2013 to conduct reviews of products under the Canadian Organic Standards.

When the organization was first founded, there were more than 40 certifiers offering certification in North America, and inputs were reviewed by a number of organizations. This process made for a confusing marketplace because different certifiers had different requirements. The national organic standards had yet to be published, and so there was no way to ensure uniform decisions about what inputs could be used on an organic farm. The CCOF had been running their own evaluation program in partnership with Oregon Tilth. But as the input market grew, CCOF, Oregon Tilth, along with the OTA, and the Organic Crop Improvement Association determined that creating a single independent organization was the only way to create a uniform standard and consolidate resources

Then, with funding from many organic organizations and businesses, OMRI was formed, and an advisory council made up of experts from among organic farmers, industry, and academia came together to create the first OMRI Generic Materials List in 1998. In creating their first list, they determined their standards and policies for reviewing new products to be included in future products lists. They used those standards to create an application process for companies who sell inputs to organic farmers. They also published their first product list in 1998. Once the organic standards were published in 2000, those regulations formed the basis of the OMRI input review process.

OMRI maintains both a products list and a generic materials list. The products list includes brand name products that have been reviewed and deemed allowed under the NOP. The generic materials list describes generic substances such as minerals or other nutrients that are compliant with the NOP. All products that have made it through the application process are given an OMRI seal to be included on labels and packaging. The application and review process includes a review of the product formulation, manufacturing processes, declarations of GMOs, radiation or nanotechnology, product labels, a lab analysis of the product, and a review fee. The applications are reviewed by an external review panel made up of experts in the specialty of the input (for example crops, livestock, processing).

An on-site visit may be necessary to complete the review. To facilitate the application process, they wrote the OMRI Policy Manual and Review Standards for Lab Analyses. They maintain a list of consultants who specialize in various input products, and who can help with the development of input products or preparing for the application. Approximately 10–15 percent of applications are either rejected because they are prohibited by the NOP or the application is withdrawn. OMRI conducts random inspections on a small fraction of listed products each year to ensure that integrity of the program is upheld.

Certifiers, businesses, and farmers rely on OMRI to review and approve products that are allowed under the NOP. Farmers, processors, and handlers must include the products in their organic system plan, and most certifiers require the products be approved by OMRI. In 2019, the OMRI products list had more than 6000 products included, and the generic materials list had over 900 substances. The lists are available online as a download and in a searchable database. Many input businesses use the OMRI seal as part of their marketing and labeling. Certifying agencies around the world also use the OMRI lists to ensure the products they are certifying will be accepted for import into the United States.

In addition to developing and maintaining the input lists, OMRI contributed to the development of the NOS and subsequent rules by providing technical reports for the NOSB between 1999 and 2002. In 2008, OMRI became ISO 17065 accredited by the Quality Assurance Division of the USDA to ensure the rigor and integrity of the program is maintained. OMRI is officially recognized by the NOP as a third-party review program that allows certifiers to contract with OMRI for material review. They resumed their role of providing technical reports to the NOSB in 2012 with permission of the NOP. OMRI is one of only three contractors allowed to bid on technical report proposals. These reports most often guide the NOP and NOSB on decisions related to the allowed substances list maintained by the NOP.

Organic Seed Alliance

The Organic Seed Alliance (OSA) has its roots in the Abundant Life Seed company founded by Forest Shomer in 1973. Shomer was a pioneer of biodiversity and heirloom seeds, who sent his first catalog out as a folded origami sheet in 1974. His company was the first in the Pacific Northwest to focus on heirloom vegetables and native plants. In 1975, he decided to transition from a for-profit company to a nonprofit foundation dedicated to seed saving and bioregional seed sources of open-pollinated varieties. His seeds were produced through a network of small-scale farmers and sold through a mail-order catalog and displayed in food cooperatives.

By the early 2000s, the organization was planning to sell the seed company portion of the organization and focus on research and education. In 2003, a fire destroyed the offices, seed archives, inventory, and the seed packaging facilities. They sold what was left of the seed company to the Territorial Seed Company and reformed the foundation into the OSA dedicated to increasing ethical seed systems that are genetically diverse and regionally adapted. OSA's programs focus on advancing seed diversity through farmer-controlled regional seed networks. They aim to reach their goals through a combination of research, education, and advocacy work.

OSA sees research as a key component to creating a vital supply of organic seeds that are adapted to specifically meet the needs of organic farmers. They emphasize collaborative research projects that bring together university researchers, farmers, and other organizations or businesses. This collaboration is key to making sure the research is truly meeting the needs of the organic sector, while growing the scientific data and field of organic seed studies. They have current projects on improving carrots and tomatoes for organic agriculture and another on identifying northern vegetable varieties for organic production. Through their work with participatory breeding, the OSA has been able to release two new seed varieties for the organic markets. The first is called Abundant Bloomsdale, a spinach variety adapted to growing in Pacific Northwest. The second is called Who Gets Kissed?—an open-pollinated variety of sweet corn adapted for organic farms in the Midwest.

One way the OSA is working to strengthen regional seed systems is to help conduct regional seed assessments. OSA brings together a range of stakeholders such as farmers, seed breeders, chefs, consumers, and local food and farming businesses to identify the specific needs of the region and strategies that will best address them in that locale. So far, they have released reports on three regional assessments: Montana Organic Seed Assessment, Pacific Northwest Organic Plant Breeding Assessment of Needs, and Southeast Organic Seed Stakeholders Survey. In 2019, they had ongoing projects in California, the Inter-mountain West, Midwest, Pacific Northwest, and Southeast United States.

The loss of seed saving skills is another area that the OSA is dedicated to addressing. OSA not only teaches farmers how to save seeds and conduct their own plant breeding, but also teaches seed saving to gardeners and hobby farmers. They offer courses throughout the United States and host a large organic seed conference every other year. Their conference focuses on growing the organic seed sector, and it attracts farmers, plant breeders, seed companies, good companies, researchers, and organic certifiers from all over the world. They pair the conference with farm tours and mini courses to expand options for learning about seed production. In addition to the courses and conference, the OSA hosts a seed internship matching service. They provide an online self-paced course and then pair interns with a host farm where the intern can learn hands-on seed saving and production techniques.

Beyond research and training, advocacy work makes up the final third of the OSA's core work. The OSA has identified the major barriers and challenges facing the organic seed sector through their State of Organic Seed project. It encompasses a large survey, meetings with focus groups, data analysis on seed research investments, and other data collection and analysis. The organic seed sector, and subsequently the entire organic sector, faces a number of challenges that the OSA believes should be addressed by advocating for better policies that support the organic seed sector. The resulting publication details the findings of the research and identifies a number of areas and actions that can improve the organic seed sector.

Using the *State of Organic Seed* report as their guide, the OSA participates in lobbying the government and educating policymakers and government staff on the benefits of supporting organic seed. They partner with the NOC, National Sustainable Agriculture Coalition, and Seeds & Breeds for the 21st Century Agriculture Coalition on policy actions and discussion at the national level. In 2019, OSA received a grant to partner with OFRF in undertaking a larger version of the national organic survey and report.

The OSA has a large collection of publications that cover seed trials, guidelines for seed saving, quick reference guides for growing vegetables, and guidelines for breeding seeds. They also produce a number of worksheets and tracking tools to support seed production and breeding trails. All of these materials are available on their website. In addition, they have created a searchable seed producers directory, and they sponsor, along with a number of other organizations, an organic seed finder database hosted by the Association of Official Seed Certifying Agencies. They have just created an online variety trial tool to simplify the variety trial planning and data analysis process. The tool will also work to aggregate data from many growers to create a larger dataset to learn from.

Organic Trade Association

The Organic Trade Association (OTA) is a membership-based trade organization representing the full organic sector comprising farmers, processors, distributors, retailers, and more. The organization's mission is to promote and protect organic and grow the sector through three main priorities: healthy markets, successful farmers, and expanded production. They accomplish this through a combination of advocacy, market analysis, international market development, and support of research and outreach programs.

The OTA actually began as the Organic Food Production Association of North America (OFPANA) in 1984. A produce distributor from the mid-Atlantic region wanted to sell organic produce from Vermont, especially cool weather crops that were hard to grow in the warm summer climates, such as spinach and lettuce. The distributor wanted the products to be certified organic and was willing to pay for farmers to get certified through a new national certification program that was intended to be producer-controlled. Twenty-three growers signed on for the first year of the deal. Around the same time, IFOAM was planning a meeting in the United States to bring together organic stakeholders and possibly form a national organization or North American association. The meeting attracted 18 stakeholders from the United States and Canada, including several people interested in a national certification program. This meeting came at a time when the numerous state and regional definitions, certification standards, and labels were confusing customers and frustrating processors, distributors, and retailers. At the meeting, the OFPANA was formed, and a second meeting planned when the organization decided to formally incorporate. A decade later, the organization changed its name to the Organic Trade Association because the original name was long and hard to remember. Like most organizations, the early years were primarily funded through personal volunteers and donors. Finally, in 1990 the organization hired Katherine DiMatteo as its first paid staff member and executive director. Katherine remained in that position until 2006.

The first priority of the new organization was to address the need for consistent national certification standards. To accomplish this, the organization decided to create a set of guidelines or a baseline that current certifiers could use to ensure their standards would be accepted in other regions. They were essentially creating an equivalency program or umbrella organization for all the current certifiers. The development of the guidelines resulted in much discussion and arguments on what standards should be included. The original goal of a national producer-controlled standard ultimately failed, but the guidelines created helped form the basis of what would become the USDA National Organic Standards. The OTA has focused considerable efforts on advocating for an organic policy at the federal level. They have formed the only organic political action coalition to support congressional candidates who commit to supporting the organic industry.

Occasionally, when the OTA is not able to achieve their goals through lobbying efforts, they have used lawsuits against the USDA. Most recently, the OTA, along with the American Society for the Prevention of Cruelty to Animals and the Animal Welfare Institute, filed a lawsuit against the USDA over its failure to implement the new organic livestock and poultry regulations. The rule was announced at the end of the Obama administration and withdrawn by the Trump Administration shortly after they took over.

As a trade organization, another priority was marketing organic to consumers and educating them on the benefits of eating organic food. The OTA is one of the main sources of data collection on the organic industry in the United States, and the organization frequently participates in international trade efforts. Their two main data collection projects result in two annual publications that the organic industry uses: The Organic Industry Survey and the U.S. Families' Organic Attitudes and Beliefs.

Over the years, there has been plenty of controversy around some of the issues the OTA supports. Many in the organic sector find the OTA to be too closely aligned with big businesses and not with organic farmers. The decision by OTA to allow nonorganic companies to join the organization and its subsequent handling of GMO-related legislation at both the state and federal levels in 2016 and the hydroponics controversy in the NOP in recent years have angered many organic farmers. There have been several public fallouts with the OTA, including in 2016, the fallout with the Organic Seed Growers and Trade Association (OSGTA) and Dr. Bronner's and in 2018 with the company Nature's Path.

The OTA recently formed a Farmers Advisory Council with representatives from organic organizations around the country. The council was created to open up dialog between small farmers and the primarily large business focused OTA. The OTA also operates the Organic Center, an independent organic science research organization that supports and partners with scientists doing research on organic food and farming. The Organic Center also focuses on communicating scientific research to the public. They often respond to negative news coverage of organic and debunk myths associated with organics.

The OTA currently represents over 9500 businesses in 50 states and includes stakeholders from all sectors of the organic industry. Its most recent efforts have focused on the creation of an organic checkoff program. OTA submitted an application to the USDA to consider a federally mandated checkoff program similar to the ones that other sectors use to fund research and promotion. The USDA did not go forward with the application because a coalition of organic farm groups opposed the program. The OTA has decided to undertake development of their own voluntary checkoff program.

Rodale Institute

J.I. Rodale, the founding father of organic agriculture in the United States, and publisher of the magazine *Organic Farming and Gardening*, established the Soil and Health Foundation in 1947 on a 63-acre farm in Emmaus, Pennsylvania, later renamed the Rodale Institute. Rodale was a publisher and passionate advocate of organic farming. He ran a publishing company called Rodale Press, which published a number of books based on research conducted at the Rodale Institute. Rodale also ran an organic certification program for a number of years.

Rodale dreamed of a demonstration and training farm to spread the idea of farming with fertile chemical-free soil. J.I. Rodale's son Robert purchased 333 acres in Kutztown, Pennsylvania, in 1971 to expand the work of the institute. The Rodale Institute is a nonprofit organization dedicated to growing the organic movement through research, farmer training, and consumer education. Their tagline, "Healthy Soil=Healthy Food=Healthy People" sums up much of the writing of founder J.I. Rodale.

The institute operates on the original 333-acre certified organic farm, which is the site of their flagship Farming Systems Trial, Vegetable Systems Trial and Hemp Research. They have three on-site laboratories, a compost yard, honeybee conservancy, an apple orchard with 30 different varieties, a pastured hog facility, demonstration gardens, and garden store. The founder's farm is home to a community-supported agriculture program, apiary, and guesthouse. More recently, the Rodale Institute has partnered with two other locations; one is an organic farm on the St. Luke's University hospital campus to provide organic food directly to the hospital food system. A second partnership was created with a local organic farm family to create a regenerative organic operation that aims to train military veterans and local community members in organic fruit and vegetable production.

In 1981, the institute started the Farming Systems Trial, now the longest running side-by-side comparison of organic and conventional grain cropping system in North America. The trial is located on 12 acres and is divided into three main systems: organic manure, organic legume, and conventional. Each system is also divided into a tillage and no-tillage system. The past 40 years have shown that organic systems are comparable and competitive with conventional systems after a five-year transition period; organic systems perform better in times of drought, use less energy, and release fewer carbon emissions. The research on these systems continues today.

Other studies pioneered by the Rodale Institute aim to address some of the most common environmental issues in agriculture and provide practical solutions to organic farmers. The Vegetable Systems Trial is also a side-by-side comparison of organic and conventional production methods started in 2017. The study is designed to examine nutrient density; soil health; drought resistance; pest, weed, and disease resistance; and profitability over 20 years. Another major study looks at the impact of organic and conventional systems on water quality. The Pastured Pork study, started in 2015, is looking for scalable models of pork production that do not rely on concentrated animal feeding operations. Other studies include hemp as a cover crop, nutrient management, composting, and pest and disease prevention.

In addition to research, the institute offers training programs through workshops, internships, webinars, and an organic farming certificate program in partnership with Delaware Valley University. They offer a special "transition to organic" course to help conventional farmers transition their farm to organic production methods. They also conduct lots of farm tours and outreach through local schools and in partnership with other organizations. They have also launched a scholarship fund to support a new generation of organic farmers and researchers.

In 2016, the Rodale Institute became the sponsor of the newly organized Organic Farmers Association (OFA). The organization was formed to provide a national voice for certified organic farmers. The organization has an active lobbyist working on behalf of the farmers. The OFA is a membership organization for certified organic farmers only and is transparent and democratic in their approach to policy positions. They also offer supporter membership and organization membership without voting rights. The organization has taken over publication of the *New Farm* magazine and offers it free to members.

Most recently, the Rodale Institute partnered with Patagonia and Dr. Bronner's to launch a new nonprofit alliance called Regenerative Organic Alliance. In 2018, the alliance began the process of developing a new certification program that goes beyond the foundation of the USDA and IFOAM standards. The Regenerative Organic Certification ran a pilot program on 22 farms in 2019 and used that experience to revise their framework. The new certification will cover soil and land management practices, animal welfare based on pasture-raised animals, and fairness for farmers and farmworkers and will always be compatible with the rules of the NOP.

The creation of the OFA and the Regenerative Organic Alliance come after many years of organic farmers feeling left out of the policymaking process at the USDA and by the lobby groups representing the organic sector.

GMO
FREE

Introduction

This chapter presents data and documents on organic food and farming. The figures illustrate some of the environmental consequences of conventional farming and some potential benefits and support for organic farming. The tables show the growth of the organic sector with lists of products in demand; states that lead the country in sales, number of farms, and acreage; and finally the organic products that top the export and import lists in the United States.

The documents section has excerpts from a number of government documents including the most important parts of the OFPA, testimony and reports to Congress on a number of key issues facing the organic sector in recent years, including testimony from two proponents of organic on the status of organic 20 years after the passing of the OFPA, and a summary of the impact of a major court case against the USDA. There is an excerpt of USDA's rationale for the withdrawal of the OLPP rule and a summary of findings from a subcommittee of the NOSB on GMO impact on organic farming.

Data

Agriculture can both cause GHG emissions and work to capture carbon and store it in the soil. Most of the emissions from

---

A farmer uses a GMO-free sign to share information about how his crops are raised. Farmers who are not certified organic cannot use the term organic to describe their crops. (Wellphotos/Dreamstime.com)

agriculture are from livestock production, tilling the soil, and producing rice.

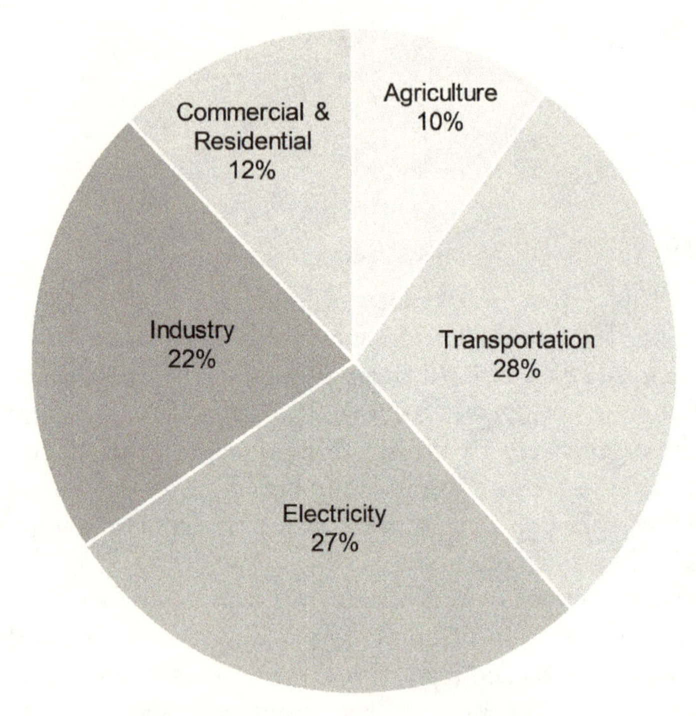

Figure 5.1. Total U.S. Greenhouse Gas Emissions by Economic Sector, 2018

Source: www.epa.gov/ghgemissions/sources-greenhouse-gas-emissions

Cover crops and rotational grazing are important agricultural practices that provide a number of environmental protections. Organic farmers use these practices twice as often as nonorganic farmers.

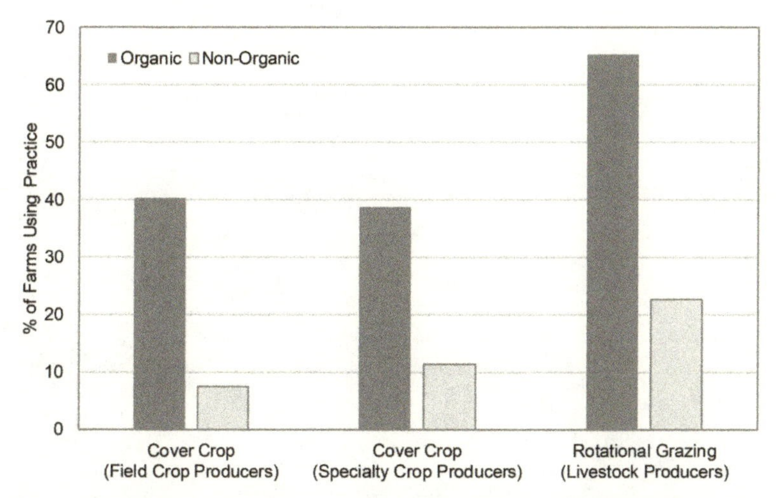

Figure 5.2. Share of Organic and Nonorganic Farms Using Conservation Practice, 2012–2014

Source: www.ers.usda.gov/data-products/chart-gallery/gallery/chart-detail/?chartId=93375

The dramatic increase in the use of GE corn and soybeans has made it very difficult for organic farmers to protect their crops from a contamination of both herbicides and treated seeds.

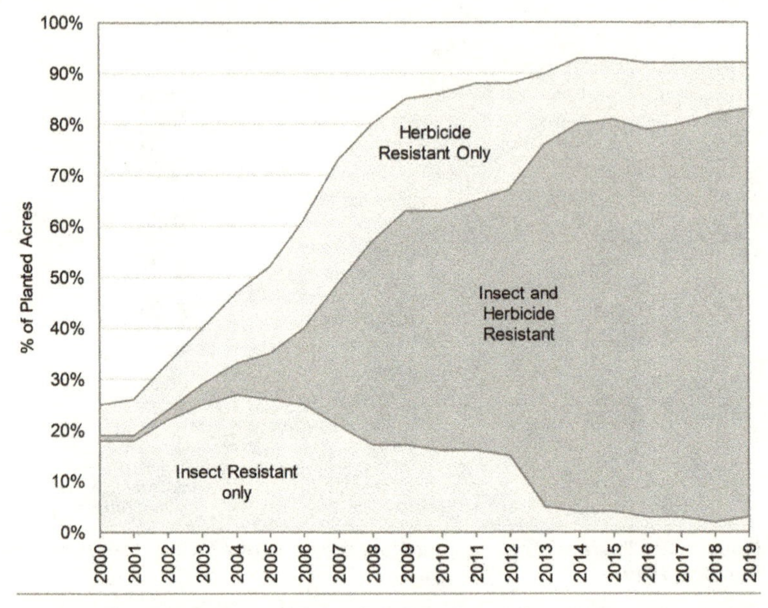

Figure 5.3. Adoption of Genetically Engineered Corn in the United States by Trait: 2000–2019

Source: www.ers.usda.gov/data-products/adoption-of-genetically-engineered-crops-in-the-us/recent-trends-in-ge-adoption.aspx

Within a decade of the introduction of commercial GE soybeans seeds, nearly all the soybeans planted in the United States were GE. That trend has continued for over 15 years.

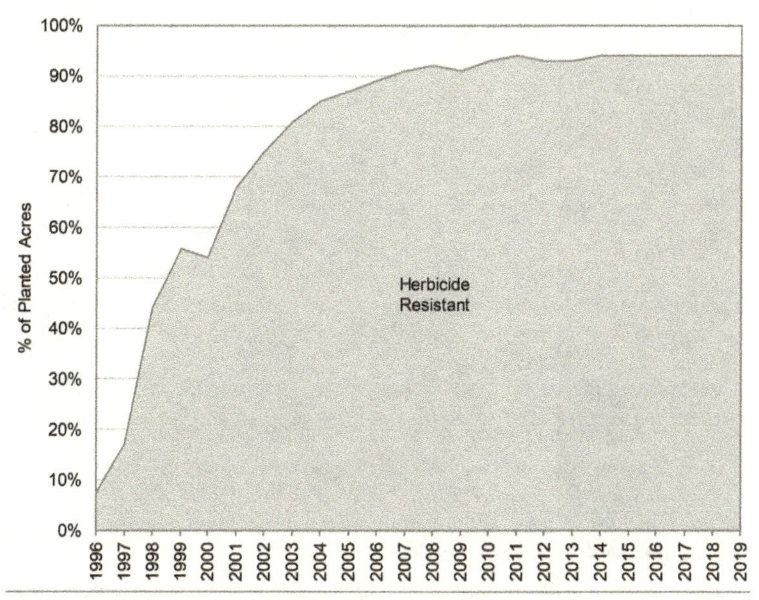

Figure 5.4. Adoption of Genetically Engineered Soybeans in the United States: 1996–2019

Source: www.ers.usda.gov/data-products/adoption-of-genetically-engineered-crops-in-the-us/recent-trends-in-ge-adoption.aspx

While overall funding for organic agriculture has increased over the years, it is nowhere close to the billions of dollars that are spent on conventional programs and research.

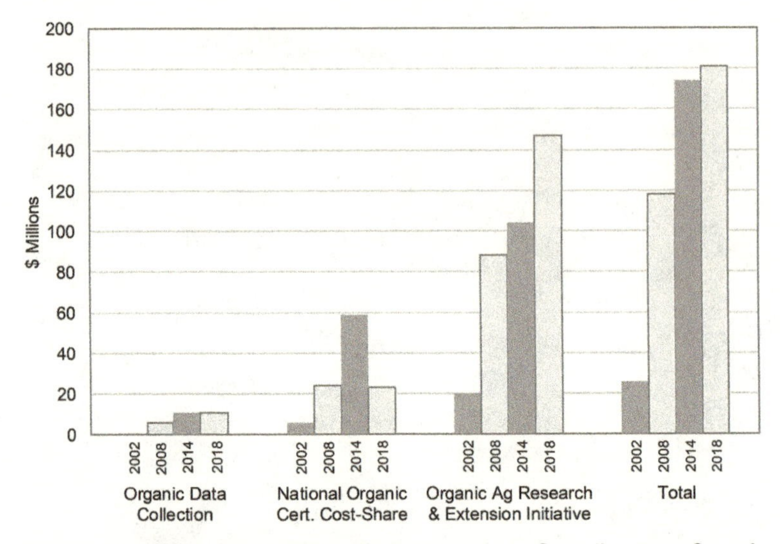

Figure 5.5. Total Inflation-Adjusted Mandatory Spending on Organic Agriculture: 2002–2018 Farm Acts

Source: www.ers.usda.gov/agriculture-improvement-act-of-2018-highlights-and-implications/organic-agriculture/

Milk at almost $1.4 billion and eggs at just over $8 million were the largest category of sales in 2016. This fits with the historical precedent. Chicken is the most-in-demand meat while apples, strawberries and lettuce topped the most-in-demand produce.

Table 5.1. Certified Organic Sales by Major Sector and Top Commodities, 2016

Commodity	Sector Total ($ million)	Commodity Sales ($ million)
Livestock and poultry products	2205	
Milk		1386
Eggs		816
Vegetables in the open	1644	
Lettuce		277
Tomatoes		175
Potatoes		151
Spinach		118
Sweet potato		101
Fruits, tree nuts, and berries	1407	
Apples		327
Strawberries		242
Grapes		218
Blueberries		101
Livestock and poultry	1157	
Broiler chickens		750
Cattle		233
Turkeys		83
Field Crops	763	
Corn for grain		164
Hay		130
Wheat		107
Soybeans		78
Tobacco		62
Nursery and floriculture	113	
Mushrooms	111	
Vegetables under protection	89	
Total	**7554**	

Source: USDA National Agriculture Statistics Service 2016 Certified Organic Survey.

California leads the country in organic sales, farms, and acreage by a wide margin. Pennsylvania and Washington have the next largest sales, but they have fewer farms and acreage than many other states. Wisconsin and New York follow California with the numbers of farms and acres. States such as Colorado and Texas have relatively few numbers of farms, but they rank higher in total acres largely due to large dairy operations.

Table 5.2. Top 10 States Ranked by Total Certified Organic Product Sales, 2016

	Total Sales ($1000)	Total Certified Farms	Total Certified Acres
California	2,889,156	2713	1,069,950
Pennsylvania	659,629	803	93,418
Washington	636,245	677	78,739
Oregon	350,896	461	194,769
Texas	297,484	217	146,801
Wisconsin	255,450	1276	219,266
New York	215,859	1059	264,385
Michigan	201,067	402	76,192
Colorado	181,297	181	176,496
North Carolina	144,917	247	31,800

Source: USDA National Agriculture Statistics Service 2016 Certified Organic Survey.

The United States exports most of its organic products to Canada and Mexico, but Japan, Taiwan, and South Korea were the next largest markets. Nearly all the large exports in 2016 were fresh fruits and vegetables.

Table 5.3. Value of U.S. Organic Exports, 2016 ($1,000)

Total Value, Selected Organic Exports (all items)	547,639
Apples	82,755
Grapes	65,795
Lettuce	56,431
Strawberries	42,374
Spinach	38,630
Carrots	30,721
Organic Tomato Sauce	22,379
Organic Coffee, Roasted (Not Decaf)	21,953
Cauliflower	21,514
Pears/Quince	18,385

(continued)

Table 5.3. Value of U.S. Organic Exports, 2016 ($1,000) (*continued*)

Total Value, Selected Organic Exports (all items)	**547,639**
Berries	16,908
Blueberries	14,380
Head Lettuce	13,988
Oranges	13,839
Lemons	13,812
Celery	12,621
Onion Sets	11,127
Broccoli	10,285
Peaches/Nectarines	9,144
Grapefruit	4,852
Cherries	4,186
Tomato	4,009
Watermelon	3,319
Peas	2,915
Potatoes	2,297
Peppers	2,254
Roma Plum Tomato	1,479
Cabbage	1,427
Beets	1,086
Cherry Tomato	966
Asparagus	916
Limes	739
Cucumbers	152

Source: USDA-FAS Trade Database.

The United States imports significantly more organic products than it exports. The top imports are a mixture of soybeans and corn used for livestock feed and products that are not commonly grown in the United States such as coffee, bananas, and olive oil. Turkey, Mexico, Italy, Peru, and Ecuador were the top suppliers of organic products to the United States.

Table 5.4. U.S. Organic Imports, 2016 ($1,000)

Total Value, Selected Organic Imports (all items)	**1,653,411**
Coffee	315,851
Soybeans (Except Seed)	250,497
Bananas	209,884
Extra Virgin Olive Oil	188,521
Yellow Dent Corn (Except Seed)	160,370
Honey	73,628
Avocado	72,667
Apples	63,676
Bell Peppers	49,435
Almonds, Shelled	39,962
Wine, Red	33,562
Green Tea	26,081
Blueberries (Cultivated)	25,400
Rice (Milled)	22,143
Wine, Sparkling	21,461
Pears	18,148
Mangoes	17,148
Wine, White	15,697
Black Tea (Tea Bags)	13,548
Durum Wheat (Except Seed)	12,677
Ginger	10,742
Flaxseed	9,013
Garlic	4,985
Virgin Olive Oil	3,316

Source: USDA-FAS Trade Database.

Documents

Excerpt from the Organic Foods Production Act of 1990

The OFPA introduced a new era in the organic movement, and nearly all of the growth of the organic sector in the United States took place after the passage of the act. It gave businesses the confidence to invest in organic products and know that organic certification would mean the same thing across the country.

PURPOSES—It is the purpose of this Act—

(1) to establish national standards governing the labeling of certain agricultural products as organically produced products;

(2) to provide consumers with reliable information concerning which products are organically produced;

(3) to assure consumers that products labeled as organically produced are not produced with, or handled with, compounds that cause adverse human health or environmental effects;

(4) to encourage environmental stewardship through the increased adoption of organic and sustainable farming methods;

(5) to assist emerging and important food industry sectors that produce, process, and market organically produced products;

(6) to provide market incentives to encourage the use of organic and sustainable farming practices;

(7) to preserve the integrity of organic food programs that have been implemented by States and encourage other States to adopt organic food programs;

(8) to facilitate interstate commerce in fresh and processed food that is organically produced; and

(9) to establish accreditation procedures for laboratories that provide certain chemical residue testing services.

SEC. 102. NATIONAL ORGANIC PRODUCTION PROGRAM

(a) IN GENERAL—The Secretary shall establish an organic certification program for producers and handlers of agricultural products that have been produced using organic methods as provided for in this title.

(b) STATE PROGRAM—In establishing the program under subsection (a)the Secretary shall permit each State to implement a State organic certification program for producers and handlers of agricultural products that have been produced using organic methods as provided for in this title.

(c) CONSULTATION—In developing the program under subsection (a), and the National List under section 111, the Secretary shall consult with any Organic Standards Board or Initial Organic Standards Board established under title II.

(d) LABELING—Each certifying agent may label agricultural products that have been produced on organically certified farms and handled through certified handling operations as "organically produced."

(e) CERTIFICATION—Each certifying agent may certify a farm or handling operation that meets the requirements of this title and the requirements of the organic certification program of the State (if applicable) as an organically certified farm or handling operation.

SEC. 110. NATIONAL STANDARDS FOR ORGANIC PRODUCTION

(a) ORGANICALLY PRODUCED PRODUCTS—To be labeled as an organically produced agricultural commodity under this title, such commodity shall meet the minimum standards established under this section.

(b) PROHIBITED PRACTICES—To be labeled as an organically produced agricultural product under this title, an agricultural commodity shall not have been produced through the use of any production practices that are prohibited under this title during the greater of 3 crop years or 3 calendar years prior to the crop year during which such agricultural product is produced.

(c) HANDLING—To be labeled as an organically produced agricultural product under this title, an agricultural commodity shall be handled in accordance with this title and the applicable State organic certification program.

(d) SYNTHETIC CHEMICALS—To be labeled as an organically produced agricultural product under this title, an agricultural commodity shall have been produced—(1) without the use of synthetic chemicals; and (2) on land that has not had synthetic chemicals applied during the greater of 3 crop years or 3 calendar years prior to the crop year during which such product is produced.

(e) SOIL QUALITY—To be labeled as an organically produced agricultural product under this title, an agricultural commodity shall not have been produced on soil or in any growing medium that has been determined by the certifying agent (after appropriate soil testing) to contain levels of chemical residue that are likely to result in unsafe residue levels in any food produced on such soil.

(f) SOIL BUILDING—To be labeled as an organically produced agricultural product under this title, an agricultural commodity shall have been produced using practices that replenish and maintain soil fertility by providing optimal conditions for soil biological activity.

(g) COMPLIANCE WITH ORGANIC FARM PLAN—To be labeled as an organically produced agricultural product under this title, an agricultural commodity shall be produced in compliance with an organic farm plan agreed to by the producer of such product and the governing State official (if applicable).

SEC. 214. ORGANIC STANDARDS BOARD

(a) DUTIES—The Organic Standards Board shall establish the proposed National List or amendments to such National List as provided in section 111 and shall submit such proposals to the Secretary.

(b) MEMBERSHIP—The Organic Standards Board shall be comprised of 15 members, of which—(1) not less than

3 shall be individuals who own or operate a certified organic farming operation; (2) not less than 3 shall be individuals who own or operate a certified organic handling operation; (3) not less than 2 shall be individuals with expertise in the areas of environmental protection and resource conservation; (4) not less than 2 shall be individuals who represent public interest or consumer interest organizations; (5) not less than 1 shall be an individual who has expertise in the field of agronomy; and (6) not less than 2 shall be individuals who have expertise in the fields of toxicology, ecology, or biochemistry.

(c) TERM—The members of the Organic Standards Board shall serve for a period of not to exceed 4 years. Such members may be reappointed to serve on the Organic Standards Board.

SEC. 215. PROPOSED NATIONAL LIST

(a) IN GENERAL—The Organic Standards Board (or the Initial Organic Standards Board) shall develop the proposed national list or proposed amendments to the National List for submission to the Secretary in accordance with section 111.

(b) CONTENT OF PROPOSED LIST—(1) ITEMIZATION OF SUBSTANCES—The proposed list referred to in subsection (a) shall contain an itemization of—(A) specific synthetic substances that may be used in the production and handling of agricultural products labeled as organically produced under title I, even though the use of such substances is prohibited elsewhere in title I; or (B) specific natural substances that shall not be used in the production and handling of agricultural products labeled as organically produced under title I, even though the use of such substances would be allowed under other provisions of title I. (2) SPECIFIC USES—An exemption under paragraph (1)(A) or prohibition under paragraph (1)(B) with respect to a specific synthetic or natural substance may be limited to a specific use of such substance in a farming or handling operation.

(c) GUIDELINES FOR PROHIBITIONS OR EXEMPTIONS—(1) EXEMPTION FOR PROHIBITED SUBSTANCES—The proposed national list may provide for the use of substances in an organic farming or handling operation that are otherwise prohibited under title I only if—(A) the Organic Standards Board determines that the use of such substances—(i) would not be harmful to human health or the environment;(ii) is necessary to the production or handling of the crop because of the unavailability of wholly natural substitute products; and (iii) is consistent with organic farming; and (B) the specific exemption is developed in accordance with this section. (2) PROHIBITION OF USE OF SPECIFIC NATURAL SUBSTANCES—The proposed national list may provide for the prohibition on the use of specific natural substances in an organic farming or handling operation that are otherwise allowed under this title only if—(A) the Organic Standards Board determines that the use of such substances—(i) would be harmful to human health or the environment; and (ii) is inconsistent with organic farming and the purposes of this title; and (B) the specific prohibition is developed in accordance with this section.

(d) REQUIREMENTS—In establishing the proposed national list or proposed amendments to the National List, the Organic Standards Board shall—(1) review available information from the Environmental Protection Agency, the National Institute of Environmental Health Studies, and such other sources as appropriate, concerning the human and environmental toxicity of substances considered for inclusion in the proposed national list; (2) work with manufacturers of substances considered for inclusion in the proposed national list to determine whether such substances contain inert materials that are synthetically produced; and (3) submit to the Secretary, along with the proposed national list or any proposed amendments, the results of the Organic Standards Board evaluation of all substances considered for inclusion in the National List.

(e) EVALUATION—In evaluating substances considered for inclusion in the proposed national list or proposed amendments to the National List, the Organic Standards Board shall determine—(1) the potential of such substances for detrimental chemical interactions with other materials used in organic farming systems; (2) the toxicity and mode of action of the substance and of its breakdown products or any contaminants, and their persistence and areas of concentration in the environment; (3) the probability of environmental contamination during manufacture, use, or misuse of such substance; (4) the effects of the substance on human health; (5) the effects of the material on biological and chemical interactions in the agroecosystem, including the physiological effects of the substance on soil organisms (including consideration of salt index and solubility), crops, and livestock;(6) the alternatives to using the substance in terms of practices or other available materials; and (7) its compatibility with a system of sustainable agriculture.

(f) PETITIONS—The Organic Standards Board shall establish procedures under which individuals may petition the Organic Standards Board for the purpose of evaluating substances for inclusion on the National List.

(g) REVIEW—(1) EVERY 4 YEARS—The Organic Standards Board shall review each substance included as an exemption or prohibition on the National List at least once during each 4-year period beginning on the date such substance was initially included on the National List or on the date of the last review of such substance under this subsection.(2) SUBMISSION TO SECRETARY—The Organic Standards Board shall submit the results of a review under paragraph (1) to the Secretary with a recommendation as to whether such substance should continue to be included on the National List.

Source: Title 7 Subtitle B Chapter 1 Subchapter M Part 205. https://www.ecfr.gov/cgi-bin/text-idx?tpl=/ecfrbrowse/Title07/7cfr205_main_02.tpl.

Excerpt from the National Organic Law at 20: Sowing Seeds for a Bright Future (2010)

Twenty years after the passage of the OFPA, the Senate committee on agriculture, nutrition, and forestry held a hearing on the state of organic farming. Several individuals were invited to present testimony in front of the hearing. This excerpt includes testimony from two of the individuals who had worked on organic farming issues for many years, including the senator who first introduced the bill.

Testimony from Senator Patrick Leahy

I am delighted to be here today to celebrate the upcoming 20th anniversary of the Organic Foods Production Act of 1990. Of course, I am joined by Senator Chambliss, a long-time friend. and just as I am a former chairman of this committee, he is a former chairman of this committee. We are being watched by former Chairman Talmadge. Chairman Lincoln, I talked with her at length about this hearing yesterday, and she has to be in Arkansas on business. But I appreciate her arranging for us to have the hearing, and she and I will be talking about the results of it when she comes back. This is an area in which she is quite interested.

We talked about the 20th anniversary of the Organic Foods Production Act. There are people here in the room who were part of that achievement. I would note especially Deputy Secretary Merrigan. She worked with us when I was chairman. And we worked with Ranking Member Lugar, Dick Lugar, to write the 1990 Farm Bill, which this was a part. At times, we felt a little bit like Sisyphus with rolling that rock, but we made it. We had a strong bipartisan coalition we put together.

But, Deputy Secretary, I just do want to acknowledge the tremendous help you gave during that. I have sometimes noted that senators are merely constitutional impediments to their staffs who do most of the work, and so I appreciate that.

But we are now looking forward to not only celebrating these 20 years, where do we go the next 20 years? Prior to the passage of the organic farm bill, the industry was growing slowly. We had farmers and consumers, retailers facing inconsistent policies and inaccurate labeling procedures across the country. And it is hard to believe today, but at the time we had 22 different states trying to manage and four separate regulations for organic foods. It made it very difficult for interstate commerce and very difficult for consumers.

The passage of OFPA brought much needed order to the industry. It gave consumers the USDA organic label, a label with real meaning. The organic law required USDA to develop a minimum national organic standard, set us on the course where we are today, certified organic farms in all 50 states, nearly 5 million acres of organic crop and pasture land, an industry with sales of more than $25 billion and growing.

I think back then when some people were asking why am I doing this organics bill; you might have a handful of farms and it is not going to go anywhere. And I told them I had listened very carefully to Vermont farmers who came to me and said, "We are willing to meet higher standards and we will do what is right, but give us some national standards so we are competing on a level playing field. As long as we follow the rules in our state, we want to know everybody is following the rules in their state."

I said at that time that the only way that this industry can grow if the standards are met and they are followed and they are enforced. Strong standards do reward farmers who play by the rules. They help consumers understand what that label means when they buy something that is USDA organic. I mean, the proof is in the pudding with a 25-billion-dollar industry that is growing. How many industries in America today can say they are growing the way this one is growing? But consumer confidence is key to the organics industry's growth. It will be the key in the future.

So we have come from those early days where everyone thought it was just a crunchy granola program. You have heard that expression. You walk in the stores, and organic foods occupy prominent shelf space in the produce and dairy aisles in the most mainstream food retailers, even big-box stores. We see the offerings, organic meats, like the delicious White Oak Pastures grass-fed beef and eggs and breads and grains, such as Annie's Cheddar Bunnies. I see we have the Cheddar Bunnies here. Beverages, even peanuts increasing with every year.

I should add, I was pleased to host Secretary Vilsack this past February at the Northeast Organic Farming Association of Vermont winter meeting in Burlington, Vermont. It is the middle of winter, Burlington, Vermont. We are a very small state. He was welcomed by more than 1,200 people who packed in to see him. No surprise, though, since Vermont leads the country in the number of organic farms on a per capita basis.

But I also recall Secretary Vilsack received an interesting organic product, a six pack of organic certified and Vermont brewed pumpkin ale. I did not ask him how he got that on the airplane afterwards, actually, nor did they consume it before they got on the airplane.

But today we have more farms and companies than ever participating in the organic sector, but we continue to experience occasional shortages of organic products when our farms are unable to simply keep up with the consumer demand. I was concerned in the past that the Department of Agriculture had not kept up with the pace of organic agriculture.

I am pleased today to see an agency that recognizes and has to support a diverse menu of options for all of American agriculture, including organic agriculture. That strong support means strong standards, and I will look forward to hearing from our witnesses today about the ongoing implementation of natural organic standards.

I am interested in the recent expansion of the national organics program at USDA. We can look back at the success of 20 years. I want to look forward to the potential success of the next 20 years, and I look forward to hearing from all of you about the potential challenges you see awaiting this young and growing industry.

I see the distinguished senator from Michigan here, Senator Stabenow. We certainly have organic farming in her state. But again, I want to thank Senator Chambliss for being here, but I especially want to thank Chairman Lincoln for letting us have this hearing.

It may be a small percentage of some of our members of the total farmland in the state, but it is growing. It is growing. And when we have all this bad news in the economy, it is kind of nice to have news about something that is working and growing, putting people to work. And I know when I walk in farmers' markets or stores and all, I see people heading to the organic food.

Testimony from Michael Sligh, founding chairman, USDA National Organic Standards Board, National Organic Coalition

Good morning. I am honored to be here on behalf of the National Organic Coalition, an alliance of farmers, ranchers, environmentalists, consumers and businesses working together to protect and enhance the integrity of organic, which is at the heart of continued consumer confidence. Thank you for this opportunity to celebrate these last 20 years of organic progress and to look to organic's bright future.

As it turns out, this is a pretty long row to hoe for the many who have been here since the beginning, but it has been a very productive one. We have made real progress, and I believe that this founding organic legislation still serves as a model on how to create a successful public-private partnership in what I might call a very vigorous hyper-participatory and transparent manner.

There, of course, have been many twists and turns, some serious failures to communicate, and even some major lapse of fair play. However, organic has survived but actually thrived against all the odds. And I believe that is because of a very unique combination of farmer innovation with marketplace entrepreneurship backed by very loyal customers and coupled by this very sound federal policy. This combination has served us very well, and we do indeed have much to be proud of as organic emerges from the margins to the mainstream.

Organic is clearly global now with standards in over 60 countries. We have witnessed over 20 years of continued growth, and the U.S. is the largest single country market in the world. Organic is even increasing yields in quality of life for some of the world's poorest farmers. To sum up, organic produces high yield and high-quality crops while reducing adverse impacts on the environment and strengthening family farms.

We also want to recognize and appreciate Congress and USDA's role which has been critical, particularly in the landmark 2008 provisions which have increased, as you have heard, many of the programs that are vital to promote organic, including the Certification Cost Share, the research, and greater access to crop insurance and conservation programs.

While I believe all of these successes are exciting, as we look ahead, I actually believe that the real potential of organic is still largely untapped. Organic is actually providing ag-based solutions to global problems of environmental degradation, climate change, food safety, declines in health and quality of life. We need to shift our thinking to publicly recognize organic not just as a marketing program but as a food system that is delivering multiple societal benefits.

So to that end, we and our organic community partners have just completed five years of dialogue developing a roadmap for organic into the future, which is the National Organic Action Plan which we will provide for the Committee. And this lays out concrete goals for the future of organic with such goals as continued doubling the number of organic products, farms and acres while ensuring fair prices to farmers, expanding research and training, expanding organic seed production, increasing local value-added processing and infrastructure, and implementing fair crop insurance and contracts for organic farms, to mention a few.

We are also very pleased that USDA and Congress has already acted on several of the key recommendations in this report, such as increasing the funding and staffing for the NOP, the pasture rule and the new policy manual, USDA's renewed commitment to oversight and enforcement, and the appointment of a USDA organic coordinator.

I also would like to point out a few of the larger societal overarching opportunities and challenges that have arisen from this report that will require your leadership and action as well. We have clearly heard from our stakeholders about this need to shift more of the responsibility for the prevention of GMO contamination back to the manufacturer. It is clear that this technology does not stay put and is threatening non-GMO markets. This must not be misunderstood as an issue between farmers or as an issue between environmentalists versus farmers, but really as an urgent need to bring overall rational market clarity and an urgent need for policy fairness, increased responsibility, and government oversight.

We also would like to highlight the food safety issues and urge that this will require a scale-appropriate, risk-based approach that is compatible with the organic practices that are already required by USDA. Organic must been seen, especially based on new research, as part of the solution to the growing food safety crisis.

I also want to highlight the concerns about seed concentration and the lack of biodiversity. As seed markets concentrate, farmers' seed costs have skyrocketed, and the diversity of public seed options have dwindled. We urgently need to reinvigorate our public plant and animal breeding capacity for a more healthy, local and nutritious diet while mitigating climate change through a more diverse and less genetically uniform agriculture. Congress has mandated this priority. We must urge USDA to fully implement this. This will be a major benefit for all farmers and society as a whole.

You have heard earlier the need for additional funding for organic research. Despite the gains in the recent Farm Bill, organic research funding still pales in the comparison. Given organic's multiple benefits to society, we think the funding level should rise to at least meet organic's current fair share.

We also need to address the need to increase access of organic foods through vulnerable populations. There is growing evidence of the public health benefits of organic, particularly for children, yet federal policy barriers are limiting these very foods to these populations. We urge that these barriers be removed. And finally, we need to better foster the next generation of organic farmers.

So in conclusion, history will not only judge us by how well we have managed our resources today but by how well we have defended the opportunities for future generations. Now is the time for us to set the course ahead for organic. Thank you very much.

Source: Hearing before the Committee on Agriculture, Nutrition, and Forestry, United States Senate, One Hundred Eleventh Congress, Second Session, September 15, 2010. Washington, DC: Government Printing Office, 2011, pp. 1–3.

Excerpt from *Harvey v. Veneman* and the National Organic Program: A Legal Analysis (2006)

The case of Harvey v. Veneman *(Secretary of Agriculture) had a profound impact on the organic farming community and changed organic legislation significantly. This case analysis provides an overview of the court case and breaks down the three main counts against the USDA and the resulting actions.*

In October 2002, Mr. Arthur Harvey filed a pro se suit against the USDA in the U.S. District Court for the District of Maine, alleging that multiple provisions of the Final Rule were inconsistent with the OFPA and the Administrative Procedures Act. The district court ruled in favor of the USDA (i.e., granted summary judgment) on all nine counts brought by Harvey. Harvey subsequently appealed the case to the First Circuit and was supported by a number of public interest groups that filed "friends of the court" or Amici Curiae briefs. The First Circuit sided with Harvey on three counts and remanded the holdings to the district court for further action. In brief, the court found that nonorganic ingredients not commercially available in organic form but used in the production of items labeled "organic" must have individual reviews in order to be placed on the National List of Allowed and Prohibited Substances; synthetic substances are barred in the processing or handling of products labeled "organic"; and dairy herds converting to organic production are not allowed to be fed feed that is only 80 percent organic for the first nine months of a one-year conversion.

The three holdings did not invalidate OFPA provisions, but rather qualified or invalidated agency regulations, thereby affecting the implementation of the National Organic Program. On June 9, 2005, the district court issued an order pursuant to the circuit court's instructions that established a two-year time frame in which the Secretary of Agriculture was to create and enforce new rules for the implementation of the National Organic Program in compliance with the circuit court's ruling. Under the order, the Secretary was to issue new regulations within a year (June 9, 2006) but has an additional year to start enforcing them (June 9, 2007). The phase-in implementation was selected by the court in an effort to prevent consumer confusion, commercial disruption, and unnecessary litigation. The rulings in Harvey and subsequent requirements for new regulations, however, were superseded in part as a result of amendments made to the OFPA by the FY2006 agriculture appropriations act (P.L. 109–97, §797). On June 7, 2006, the

USDA published revised final rules based on Harvey and the amended OFPA.

The amendments made in the appropriations measure address many of the legal concerns (e.g., lack of authority for agency action) observed by the First Circuit. The following paragraphs examine each holding where the court determined that a provision of the Final Rule was inconsistent with the OFPA and then discuss the effect of the applicable provisions from the appropriations act. Each section ends with the USDA's latest regulatory action.

Count One: Alleged Exemption for Nonorganic Products Not Commercially Available

Court Action

Plaintiff challenged the portion 7 C.F.R. §205.606 which permits the introduction of nonorganically produced agricultural products as ingredients in, or as substances on, processed products labeled as "organic" when the specified product is not commercially available in organic form. The regulation lists five specific products—Cornstarch, Gums, Kelp, Lecithin, and Pectin—and also allows for any other nonorganically produced agricultural product when the product is not commercially available in organic form. The OFPA, however, requires all specific exemptions to the Act's prohibition on nonorganic substances to be placed on the National List following notice and comment and periodic review. Harvey claimed that §205.606 provided a blanket exemption to the OFPA's review requirements and allowed ad hoc decisions to be made regarding the use of synthetic substances. The USDA, on the other hand, maintained that the regulation does not establish a blanket exemption, but rather, only permits the use of the five products specifically listed in the section. The court found the USDA's interpretation plausible; however, because the district court did not clarify the regulation's meaning, the circuit court also found Harvey's interpretation potentially credible. Accordingly, the court remanded the count to the district court for entry of a declaratory judgment that would interpret the regulation in a manner consistent with the National List requirements of the OFPA.

A declaratory judgment stating that §205.606 does not establish a blanket exemption to the National List requirements in statute for nonorganic agricultural products that are not commercially available was issued on June 9, 2005. The USDA, in compliance with the order, issued a Notice in the Federal Register clarifying the meaning of the regulation on July 1, 2005. However, because of the potential for confusion, the order states that the clarified meaning of §205.606 will not become effective and enforceable until two years from the date of the judgment (June 9, 2007).

Congressional Action

In the FY2006 agriculture appropriations act, Congress amended 7 U.S.C. §6517(d)—titled "Procedure for Establishing a National List"—to authorize the Secretary of the USDA to develop emergency procedures for designating agricultural products that are commercially unavailable in organic form for placement on the National List for a period of no longer than 12 months. The amendment does not define what an "emergency procedure" would entail; thus, the Secretary would appear to have the authority to describe the term's parameters and to select the substances subject to it. While this amendment creates an expedited petition process for commercially unavailable organic agricultural products, it does not appear to alter the ruling described above.

Administrative Action

The new rule published on June 7, 2006, did not clarify the conditions of "emergency procedure." However, it clearly restated that the five listed substances were the only nonorganically produced products that could be used as ingredients in organic products, subject to agency restriction when that ingredient is not commercially available in organic form.

Count Three: Use of Synthetic Substances in Processing

Court Action

Plaintiff challenged 7 C.F.R. §205.600(b) and the portion of §205.605(b) that permits synthetic substances as ingredients in, or as substances on, processed products labeled as "organic." Section 205.600(b) provides that synthetic substances may be used "as a processing aid or adjuvant" if they meet six criteria; §205.605(b) lists 38 synthetic substances specifically allowed in or on processed products labeled as "organic." The court found that 7 U.S.C. §6510(a)(1) and §6517(c)(B)(iii) forbid the use of synthetic substances during the processing or handling of a product, unless otherwise required by law. The court noted that the OFPA contemplates the use of certain synthetic substances during the production or growing of organic products, but not during the handling or processing stages. By allowing the use of certain synthetic substances "as processing aids," the court concluded that the regulations contravened the plain language of the OFPA. The circuit court reversed the district court's grant of summary judgment and remanded the count to the district court for entry of summary judgment in Harvey's favor. On remand, the district court ordered the Secretary of the USDA to publish new rules implementing the circuit court's judgment within one year of the date of the judgment (June 9, 2006), but allowed the Secretary to exempt nonconforming products placed in commerce as "organic" for up to two years after the date of the judgment (June 9, 2007).

Congressional Action

The FY2006 agriculture appropriations act amended §6510(a)(1) and strikes §6517(c)(B)(iii)—provisions that the First Circuit relied upon to emphasize that synthetics were not allowed during the processing or handling of a product. Before the amendment, §6510(a)(1) barred a person on a handling operation from adding any synthetic ingredient during the processing or postharvest handling of a covered product. The amendment added the phrase "not appearing on the National List" after "ingredient," thereby apparently allowing the use of synthetics on the National List during processing or postharvest handling of a covered product. Section 6517(c) establishes guidelines for placing substances on the National List and in subsection (B) sets forth specific requirements with regard to the types of substances that may be exempted for use in production and handling. Specifically, subpart (iii) of §6517(c)(B) states that the substance "is used in handling and is non-synthetic but is not organically produced" (emphasis added). This provision, which the court noted "specifically requires the exempted substances be nonsynthetic [sic]," was deleted by the amendment. As there no longer appears to be any general prohibition (though there are other requirements that must be met) against the placement of synthetics on the National List for use during the processing or handling of a covered product, the First Circuit's ruling in count three is likely moot.

Administrative Action

The USDA determined that there was no need to revise §205.600(b) and §205.605(b) because Congress sufficiently addressed the contradiction and approved the necessary legislative changes.

Count Seven: Conversion of Dairy Herds to Organic Production

Court Action

Plaintiff challenged the Final Rule's exception to the OFPA's requirements for dairy herds being converted to organic production. Pursuant to 7 U.S.C. §6509(e)(2), a dairy animal whose milk or milk products will be sold or labeled as organically produced must be raised and handled in accordance with the OFPA for not less than the 12-month period immediately prior to the sale of such milk or milk products. Section §205.236(a)(2) of the Final Rule, however, allows whole dairy herds transitioning to organic production to use 80% organic feed for the first nine months and 100% organic feed for the final three months (i.e., "80–20" rule). The court found the OFPA's requirement for a single type of organic handling for twelve months and the Final Rule's bifurcated approach in direct conflict. The court determined that nothing in the OFPA's plain language permits the creation of an "'exception' permitting a more lenient phased conversion process for entire dairy herds," and consequently, found the regulation invalid. The circuit court reversed the district court's grant of summary judgment and remanded the count to the district court for entry of summary judgment in Harvey's favor. On remand, the district court ordered the USDA to promulgate regulations implementing the circuit court's decision within one year of the date of the judgment (June 9, 2006) and to start enforcement by June 9, 2007.

Congressional Action

In the FY2006 agriculture appropriations act, Congress amended 7 U.S.C. §6509(e)(2) by adding an exception to the general feeding requirement listed in the provision (i.e., raised and handled in accordance with the OFPA for not less than the 12-month period immediately prior to sale). The new provision, titled "Transition Guideline," allows crops and forage from land included in the organic system plan of a dairy farm that is in the third year of organic management to be consumed by the dairy animals of the farm during the 12-month period immediately prior to the sale of the organic milk or milk products. Generally, crops or forage intended to be sold or labeled as "organic" cannot have prohibited substances applied to them for the three years immediately preceding harvest of the crop. Accordingly, while this amendment allows feed for dairy animals to come from land that is still transitioning to "organic" status, it would not appear to allow dairy cows to be fed prohibited substances or genetically modified organisms. Congress' amendment to §6509 likely made the court's ruling in count seven moot.

Administrative Action

The Secretary revised 7 C.F.R. §205.236 to create two exceptions to the general rule that milk labeled as "organic" must come from cows under continuous organic management for no less than 12 months. First, animals may consume crops and forage from the producer's land that is in the third year of organic management (i.e., the transition guideline). Second, producers converting entire herds to organic production who were still using the "80–20" feed rule before the publication of the new regulation may continue to do so, provided that no milk may be labeled as "organic" by this method after June 9, 2007. This exception allows a period of transition to occur in accordance with the court's order for enforcement of new regulations by the same date.

Source: Congressional Research Service Report, September 26, 2006, Order Code RS22318.

Excerpts from Organic Livestock and Poultry Practices Final Rule; Withdrawal (2018)

In 2011, the NOSB recommended additional rules to clarify outdoor access and other practices required to maintain organic certification. The USDA finally proposed a rule in 2016 after significant lobbying from both sides of the issue. A final rule was published in 2017, but rescinded in 2018. The excerpt below is the USDA's explanation for the withdrawal of the final rule.

In May 2018, the USDA withdrew the Organic Livestock and Poultry Practices (OLPP) final rule published on January 19, 2017. The rule would have increased federal regulation of livestock and poultry for certified organic producers and handlers. The USDA received 72,000 comments on the proposal to withdraw the OLPP final rule with over 63,000 opposing the withdrawal.

1. ANALYSIS OF ITS AUTHORITY UNDER THE OFPA TO ISSUE STAND-ALONE ANIMAL WELFARE REGULATIONS

The OLPP final rule consisted, in large part, of rules clarifying how producers and handlers participating in the National Organic Program must treat livestock and poultry to ensure their wellbeing (82 FR 7042). AMS is withdrawing the OLPP final rule because it now believes OFPA does not authorize the animal welfare provisions of the OLPP final rule. Rather, the agency's current reading of the statute, given the relevant language and context, is that OFPA's reference in 7 U.S.C. 6509(d)(2) to additional regulatory standards "for the care" of organically produced livestock does not encompass stand-alone concerns about animal welfare, but rather is limited to practices that are similar to those specified by Congress in the statute and necessary to meet congressional objectives outlined in 7 U.S.C. 6501.

USDA believes that the Department's power to act and how it may act are authoritatively prescribed by statutory language and context; USDA believes that it may not lawfully regulate outside the boundaries of legislative text. Therefore, in considering the scope of its lawful authority, USDA believes the threshold question should be whether Congress has authorized the proposed action. If a statute is silent or ambiguous with respect to a specific issue, then USDA believes that its interpretation is entitled to deference and the question becomes simply whether USDA's action is based on a permissible statutory construction.

The OLPP final rule is a broadly prescriptive animal welfare regulation (82 FR 7042, 7074, 7082). USDA's general OFPA implementing authority was used as justification for the OLPP final rule, which cited 7 U.S.C. 6509(g) as "convey(ing) the intent for the USDA to develop more specific standards.. .." (82 FR 7043), and 7 U.S.C. 6509(d)(2) as authorizing regulations for animal "wellbeing" and the "care of livestock" (82 FR 7042, 7074, 7082).

But nothing in section 6509 authorizes the broadly prescriptive, stand-alone animal welfare regulations contained in the OLPP final rule. Rather, section 6509 outlines discrete aspects of animal production practices and materials relevant to organic certification: sources of breeder stock, livestock feed, use of hormones and growth promoters, animal health care, and recordkeeping. While subsection 6509(d)(2) authorizes promulgation of additional standards for the "care" of livestock, that provision is not free-standing authority for AMS to adopt any regulation conceivably related to animal "care"; rather, standards promulgated under that authority must be relevant to "ensur[ing] that [organic] livestock is organically produced" 7 U.S.C. 6509(d)(2). Similarly, section 6509(g) is not open-ended authority to regulate any and all aspects of livestock production; rather, it authorizes AMS to promulgate regulations to "guide the implementation of the standards for livestock products provided under this section" (emphasis added); in other words, standards relevant to and necessitated by the expressed purposes of Congress in enacting the OFPA. Thus, standards promulgated pursuant to section 6509(d)(2) and section 6509(g) must be relevant to ensuring that livestock is "organically produced."

Although Congress did not define the term "organically produced" in the OFPA, the Cambridge Dictionary defines "organic" as "not using artificial chemicals in the growing of plans and animals for food and other products." Merriam-Webster defines "organic" as "of, relating to, yielding, or involving the use of food produced with the use of feed or fertilizer of plant or animal origin without employment of chemically formulated fertilizers, growth stimulants, antibiotics, or pesticides" (emphasis added) https://www.merriam-webster.com/dictionary/organic. The surrounding provisions in section 6509 demonstrate that Congress had a similar understanding of the term "organic." For example, subsection 6509(d)(2)'s authority for promulgation of additional standards governing animal "care" is contained within a subsection entitled "Health care" and follows a list of three specifically prohibited health care practices that each relate to ingestion or administration of chemical, synthetic, or nonnaturally occurring substances: Use of subtherapeutic doses of antibiotics; routine use of synthetic internal parasiticides; and administration of medication, other than vaccines, absent illness. AMS believes these prohibited practices—all of which relate to ingestion of chemical, artificial, or nonorganic substances—are representative of the types of practices and standards that Congress intended to limit exposure of animals to nonorganic substances and thus "ensure that [organic] livestock is organically produced." Thus, the authority provided by section 6509(d)(2) does not extend to any and all aspects of animal "care"; it is limited to those aspects of animal care that are similar to the examples provided in the statue and relate to ingestion or administration of nonorganic substances, thus tracking the purposes of the OFPA.

Reading this language in context, AMS now believes that the authority granted in section 6509(d)(2) and Start Printed Page 10777section 6509(g) for the Secretary to issue additional regulations fairly extends only to those aspects of animal care that are similar to those described in section 6509(d)(1)—i.e., relate to the ingestion or administration of nonorganic substances, thus tracking the purposes of the OFPA—and that are shown to be necessary to meet the congressional objectives specified in 7 U.S.C. 6501.

AMS finds that its rulemaking authority in section 6509(d)(2) should not be construed in isolation, but rather should be interpreted in light of section 6509(d)(1) and section 6509(g). Furthermore, AMS believes that a decision to withdraw the OLPP final rule based on § 6509's language, titles, and position within Chapter 94 of Title 7 of the United States Code; 3 controlling Supreme Court authorities; and general USDA regulatory policy, would be a permissible statutory construction.

Source: Docket Number: AMS-NOP-15-0012; NOP-15-06PR More information on the OLPP final rule is available in the March 12, 2018, Federal Register.

Excerpt from NOSB Materials/GMO Subcommittee Proposal Report on Progress to Prevent GMO Incursion into Organic (2016)

GE seeds and GMO products are a significant threat to the integrity of the organic food supply, and yet, very little is done to protect organic farmers from contamination. There are some in the organic industry that would like to include GMO food in organic certification. A NOSB committee was tasked with looking at the impact of such a large regulation change. These are some of their main findings.

Five years ago, and at the request of the wider organic community, the NOSB accepted the responsibility for making recommendations to the NOP to address issues related to "excluded methods" as defined in the Federal Organic rule; specifically, to ensure that Genetically Modified Organisms (GMOs) are prohibited in organic production and handling.

The issues around GMOs in organic agriculture are complex and will require long-term efforts. Therefore, we believe that advising you on our efforts to date is worthwhile as we move towards a new Administration. In the NOSB's initial letter to you in March 2012 we stated, "We would like to open the door to continued dialogue with the USDA so that the responsibility to prevent GMO contamination of organics is shared by those who develop, use, and regulate this technology. USDA actions are critical to the integrity of the organic seal and consumer confidence." This report elaborates on the progress by the NOSB with regard to this shared responsibility.

The Public's Message Is Clear

The NOSB has the unique opportunity of having direct access to public comment prior to each of our twice-yearly board meetings, and one message has consistently, repeatedly and abundantly been made clear: consumers across the country have expectations there will be no GMOs in their organic food. The risk to the integrity of organic agriculture is significant, and seed producers, growers, processors and handlers are all potentially impacted by the risk of incursion of GMOs into an organic supply chain.

NOSB Actions to Date

To address public concerns, five years ago the NOSB established an ad hoc Committee on GMOs, which has since been incorporated into the standing NOSB's Materials subcommittee. Since 2012 the NOSB has undertaken the following:

- Developed a mission statement that states that we accept responsibility for making recommendations that aim to keep GMOs out of organic agricultural products, and that we will provide leadership in clarifying the rule regarding excluded methods.
- In 2012, began work on the issues of keeping seed stocks free from GMO incursion. The work included multiple discussion documents open for public comment, an expert discussion panel at the Spring 2015 NOSB meeting, and an update report on the work of the subcommittee. Most recently, the NOSB has requested an ongoing stakeholder task force to continue working on details of data collection and threshold identification of needs. While there is still much work to be done, we can say with confidence that the organic industry has reached consensus on several key points:
- Seed is an important place to start to make sure that GMOs do not enter the organic agroecosystem. More data is needed on the sources of how GMOs can contaminate organic systems; whether it enters through seed, through pollen, or through post-harvest handling activities.
- The organic industry alone should not bear the costs of genetic trespass and incursion. The responsibility should particularly lay with the developers of these technologies that trespass on the integrity of organic production.
- In 2013 we started to examine the definition of excluded methods in the Federal Rule to see how it could be

strengthened. The definition had been developed in 1995, and many new technologies and approaches have been adopted since then. After two discussion documents and an initial proposal, the NOSB will vote on a final document this fall. This document proposes guidance on additional definitions to supplement the one in the Rule, as well as guidance on principles and criteria for the NOSB to use when reviewing future biotechnologies.

- A comprehensive recommendation was passed unanimously by the NOSB in October 2015 on Prevention Strategies for Excluded Methods. This included best management practices (BMPs) to ensure the integrity of every step of organic production and handling.

These activities have kept the topic of genetic engineering on every NOSB agenda for the last five years and have given organic stakeholders ample opportunity to comment on these issues. Again, the message has been clear: organic consumers do not want GMOs in their food, and organic farmers do not want GMO incursion into their fields or the toxic pesticides and herbicides that the use of GMOs proliferates.

USDA Leadership Is Critical

Recognition of the potential for unfair burden to be placed on non-biotech farming systems was clear in the mandate from your office to the USDA Advisory Committee on Biotechnology and 21st Century Agriculture (AC21), as evidenced by their November 2012 report, "Enhancing Coexistence: A Report of the AC21 to the Secretary of Agriculture."

In the introduction to that report, AC21's mandate was to answer the following questions:

1. What types of compensation mechanisms, if any, would be appropriate to address economic losses by farmers in which the value of their crops is reduced by unintended presence of genetically engineered (GE) material(s)?

2. What would be necessary to implement such mechanisms? That is, what would be the eligibility standard for a loss and what tools and triggers (e.g., tolerances, testing protocols, etc.) would be needed to verify and measure such losses and determine if claims are compensable?
3. In addition to the above, what other actions would be appropriate to bolster or facilitate coexistence among different agricultural production systems in the United States?

In one of the report's conclusions, it states: "In its examination of the charge provided by the Secretary, the members of the AC21 have concluded that the responses to all three elements of that charge are linked. No member of the AC21 believes that simply putting in place a compensation mechanism to address economic losses to farmers arising from unintended presence of GE or other material would completely eliminate such unintended presence and strengthen relations between neighboring farmers."

As evidenced by this report, the issues of coexistence are clearly complex. The NOSB urges the Administration to continue to show leadership by facilitating further discussion on these issues. In particular, many organic stakeholders believe the USDA's actions on genetically engineered crops have been insufficient to protect the organic industry. The NOSB urges you to prioritize the protection of the integrity of the organic industry, which as of 2015 has reached over $43.3 billion in annual domestic sales (Organic Trade Association survey).

Specifically, the NOSB urges you and your agency to:

- Develop policies to address shared responsibilities for GMO contamination.
- Strengthen farming best management practices guidance to prevent incursion of biotech seeds, pollen and products into conventionally and organically managed acreages.
- Support funding for research and data collection on threshold testing of organic and non-GMO seeds.

The NOSB appreciates that a cornerstone of your administration has been the growth of Organic agriculture. We urge you to continue to champion organic integrity through support of concrete steps—including vigorous, targeted regulatory action—to ensure the concept of coexistence is implemented in an effective, balanced and fair manner.

Source: www.ams.usda.gov/sites/default/files/media/MSNOSBReporttoSecyNov2016v2.pdf.

Introduction

Most Americans have at least heard of organic agriculture, but few can actually define what it is and know how to find good information about it. There is a wide array of sources on organic food and farming available online, in books, reports, and print media, but because the topic can be controversial, there are many unreliable sources of information. The books, articles, reports, films, and websites listed here provide a good starting point to explore the history and current issues prevalent in the organic sector. The focus here is on more recent books and articles with the exception of a few classic or foundational resources. A number of review articles are included because they are particularly useful in summarizing the topic and referencing related sources. A brief summary of the main topics and overview of each source is included.

Books

Barker, Allen V. 2016. *Science and Technology of Organic Farming*. Boca Raton, FL: CRC Press.

There are very few crop science books focused entirely on organic production methods. This book fills that gap by providing an informative and practical resource on soil fertility, plant pathology, entomology, and yields. There

---

Organic green onions growing in a greenhouse. Farmers often use greenhouses to extend the growing season. (Ligita Kluga/Dreamstime.com)

are detailed explanations of all the required plant nutrients and how to achieve the right balance of each. It contains over 50 illustrations and is designed to be a practical guide for those who have some background with soil and plant science. It is also a good resource for making a solid scientific case for the validity of organic farming techniques.

Barton, Gregory A. 2018. *The Global History of Organic Farming.* Oxford: Oxford University Press.

Barton is a historian who writes about environmental history and trade. In this book, he delves into the origins of the organic movement. He puts particular emphasis on the how the organic movement has engaged with scientific research and environmentalism over time. He dedicates several chapters to the role of Sir Albert Howard and both his first and second wives and their collective role in advocating for organic farming methods rather than industrial chemical farming practices.

Bellon, Stéphane, and Servane Penvern, eds. 2014. *Organic Farming, Prototype for Sustainable Agricultures.* Dordrecht; New York: Springer.

The editors of this book have brought together a wide range of experts to write chapters that illustrate how organic farming is the ideal model for sustainable agriculture. Each of the 25 chapters addresses a specific way in which organic agriculture can contribute to solutions ranging from technological approaches to policy and marketing programs that support sustainable agriculture goals. The chapters can be fairly technical or theoretical, so it is geared more toward an academic audience.

Carlisle, Liz. 2015. *Lentil Underground: Renegade Farmers and the Future of Food in America.* New York: Gotham Books.

This excellent book reads like a novel, but it's filled with detailed information about the growth of the organic sector. Carlisle traces the birth and growth of an entire industry of legume growers in Montana by following the history of one company. The book tells the story of a small group of farmers who came together and created a thriving business on nothing but hard work and chance. Along the way, the book also describes the growth of the organic sector, what leads various farmers to leave chemical farming behind, and the obstacles one faces when learning to farm organically. You also get behind-the-scenes glimpses of the development of the Montana Organic certification program. It highlights the different types of financial struggles that conventional and organic farmers face and the possibility organic markets present to those willing to take the risk.

Chrzan, Janet, and Jacqueline A. Ricotta, eds. 2019. *Organic Food, Farming, and Culture*. New York: Bloomsbury Academic.

This book provides a straightforward overview of the history, practices, and supply chains involved in the organic farming industry. It takes a global view and includes a number of case studies and profiles from around the world. It is written by a collection of academic authors who provide good background information on some of the major issues in organic food and farming.

Clark, Lisa. 2015. *Changing Politics of Organic Food in North America*. Northampton, MA: Edward Elgar Publishing Ltd.

The dichotomy of values versus industry is well established in the organic sector, and this book is one of many that try to understand that influence on the structures that hold the sector together. This book looks at both the associations and the businesses that shaped the institutional structures in place in North America. There are sections that cover the creation of the regulations, markets, trade agreements, and the definitions underpinning the entire movement. The book is scholarly, but well written and easy to read.

Duram, Leslie A. 2005. *Good Growing: Why Organic Farming Works*. Lincoln: University of Nebraska Press.

This was one of the first books to be written on organic agriculture that explored the issues from the perspective of the farmers. It covers the experiences of farmers in five different growing regions across the United States. In-depth interviews allow for detailed examples of what motivates organic farmers, how they find their information about organic practices, what influences their decision-making, and how they practically make an organic farm work. The book also provides good context of the social, political, and economic frameworks that those farmers are operating under and how those frameworks influence their choices. Plenty of attention is paid to the available research and information on organic farming, or lack thereof, and solutions for improving the success of organic farmers.

Fitzmaurice, Connor, J., and Brian J. Gareau. 2016. *Organic Futures: Struggling for Sustainability on the Small Farm*. New Haven: Yale University Press.

This well-written and researched book provides an inside look at small-scale organic farmers, their markets, and their personal values. The first half of the book provides a good overview of the organic movement in the United States and sets off the context in which small organic farmers are now operating. The book contains insights into the challenges farmers face and how they overcome those challenges to create successful operations. The book concludes with a number of suggestions on how to support small farmers and create strong regional food systems.

Francis, Charles, ed. 2009. *Organic Farming: The Ecological System*. Agron. Monogr. 54. Madison: CSSA, SSSA.

This scholarly book, published by the American Society for Agronomy, draws on highly regarded researchers and scientists to provide a practical and comprehensive resource on organic farming practices. It briefly covers the history and legality of organic farming and then offers several in-depth chapters about specific organic farming management practices, such as soil fertility, weed control, and a number of different systems designs that are important for organic farming. It finishes with a discussion about the importance of marketing, education, and research to the future of organic production.

Fromartz, Samuel. 2007. *Organic, Inc.: Natural Foods and How They Grew.* New York: HMH.

Written by a business writer, this book examines the growth of the organic food sector in the United States. The book covers all the issues of scale and growth in the organic sector, using a conversational narrative. Each chapter profiles businesses and individuals with in-depth interviews that illuminate the issues. The author highlights the successful growth of a number of organic companies and their often humble origins.

Guthman, Julie. 2014. *Agrarian Dreams: The Paradox of Organic Farming in California.* Second edition. California Studies in Critical Human Geography 11. Oakland: University of California Press.

California is the top producer of organic fruits and vegetables in the United States. It was one of the earliest states to enact organic legislation and had numerous organic advocates working on organic issues since before the passage of the OFPA. This book covers the growth of organic agriculture in California in relation to the regional history. It also looks at the evolution of organic regulation and how it changes the local industry. This is one of the few books that looks at labor issues on organic farms and how that ties into the balance of organic ideals versus organic in practice. Guthman does an excellent job of portraying the rise of industrial organic farms and the shifts that come from becoming a regulated industry. She illustrates her points with examples drawn from numerous interviews and case studies.

Haedicke, Michael A. 2016. *Organizing Organic: Conflict and Compromise in an Emerging Market.* Stanford, CA: Stanford University Press.

Is organic an industry or a social movement? This book focuses on the identity crisis that organic has faced over the course of its exponential growth. Haedicke draws on interviews and archival data to explore the conflicts, personal and organizational identities, and the resulting institutional or organizational structures that have been created from these two narratives. The book draws on a number of theoretical frameworks and is deeply rooted in scholarly literature and therefore a more difficult read.

Howard, Albert. 2006. *The Soil and Health: A Study of Organic Agriculture. Culture of the Land: A Series in the New Agrarianism.* Lexington: University Press of Kentucky.

This is a new release of a classic book written by the founding father of organic agriculture. This book and *An Agriculture Testament* are the seminal works of Sir Albert Howard, and they have influenced generations of organic farmers and researchers. *The Soil and Health* introduces the importance of soil health to all aspects of farming and provides detailed explanations of Sir Albert Howard's research and the resulting practices he developed with the help of those working at his research station in India, including his wife, Gabrielle. The wonderful author and poet, Wendell Berry, introduces this text with a look at its relevance in today's context.

Howard, Albert, Sir. 1940. *An Agricultural Testament.* London; New York: Oxford University Press.

In *An Agricultural Testament*, Howard describes his Indore composting method and how it impacts soil fertility in great detail. He also includes a section that criticizes agricultural research occurring at the time and presents a model for more effective methods. This classic book is still highly relevant and useful in understanding basic principles of compositing, soil fertility, and approaches to agricultural research.

Laufer, Peter. 2014. *Organic: A Journalist's Quest to Discover the Truth behind Food Labeling*. Guilford, CT: Lyons Press, imprint of Globe Pequot Press.

Laufer explores the issue of organic fraud by tracing the source of his organic food to the origins. Along the way, he files a formal complaint with the USDA to expose the limitations of organic certification, especially when it comes to imported food. The book is personal, interesting, and easy to read, but it's still packed with valuable information about how the organic import business works and the challenges facing retailers in obtaining organic food.

Lockeretz, William, ed. 2011. *Organic Farming: An International History*. Cambridge, MA: CABI.

This book provides an excellent overview of history of organic policy, research, and markets from around the world. Each chapter is written by a leading scholar or director of a nonprofit organization familiar with the aspects and status of organic in their own country. The book is particularly useful for comparing the impacts of different programs around the world and finding further resources on organic activities in a particular region.

Mosier, Samantha L. 2017. *Creating Organic Standards in U.S. States: The Diffusion of State Organic Food and Agriculture Legislation, 1976–2010*. Lanham, MD: Lexington Books.

This is a scholarly book that covers the history of organic regulation from a state policy perspective. It covers the regional differences in politics and socioeconomics that influenced the adoption of organic regulation. The book includes case studies of the two states that were early adopters and one late adopter of organic legislation. While it uses quite a bit of theoretical language and advanced statistics, it contains some otherwise hard-to-find information about the early regulations.

Obach, Brian. 2015. *Organic Struggle: The Movement for Sustainable Agriculture in the United States.* Cambridge, MA: MIT Press.

An incredibly detailed and rich history of the organic movement in the United States, this book lays the groundwork for understanding how corporate organic agriculture was created. The book presents a wide array of actors and ideologies that form the spectrum of organic markets and values at play. Obach adeptly shows how the evolution of the organic sector naturally led to certain unavoidable outcomes and what was gained or lost in the process. He argues that policy and not market forces will be the limiting factor in increasing the proportion of organic food grown in the United States.

O'Sullivan, Robin. 2015. *American Organic: A Cultural History of Farming. Gardening, Shopping, and Eating.* Lawrence: University Press of Kansas.

This book follows the growth of the organic movement with a cultural lens. O'Sullivan explores the social changes that have occurred along with the evolution of the organic sector and the role of early publications on organic agriculture in creating that change. It begins with the cultural ideals of the organic movements' origins and finishes with the identity construction of people who buy organic food. O'Sullivan draws on three main areas of scholarship: environmental history, consumer studies, and food studies.

Reed, Matthew. 2010. *Rebels for the Soil: The Rise of the Global Organic Food and Farming Movement.* New York: Taylor & Francis.

This book helps readers understand the social movement aspects of organic food and farming. It describes several distinct periods of time that shift the movement forward while focusing on the activities in mainly English-speaking countries. Reed primarily focuses on the efforts of those in the United Kingdom and, to a lesser extent, the United States. There is an entire chapter dedicated to the impact of GM crops on the organic movement and a suggestion of how the movement may evolve in the future. This is a scholarly book driven by social movements theory, but it contains many useful explanations of the trajectory of organic.

Ronald, Pamela C., and Raoul W. Adamchak. 2011. *Tomorrow's Table: Organic Farming, Genetics, and the Future of Food.* New York: Oxford University Press.

No issue is more controversial in the history of the organic sector than the use of GMOs in organic agriculture. This book explores the possibility of including GMOs in organic production from the point of view of a married couple. Pamela is a plant scientist and Raoul is an organic farmer. They take turns covering the basics of both genetic engineering and organic production methods and the controversies around them in clear and engaging language. They move the argument away from the current fight over the ways GMO crops are used now to discuss the possibilities of how it could be used to benefit farmers and consumers given more research and open dialog between farmers and researchers.

Articles

Aschemann-Witzel, Jessica, and Stephan Zielke. 2017. "Can't Buy Me Green? A Review of Consumer Perceptions of and Behavior toward the Price of Organic Food." *Journal of Consumer Affairs* 51 (1): 211–51.

This article explores the relationship between price, perceptions, and knowledge of price and purchasing of organic food. The authors drew on research literature between 2000 and 2014 in North America and Europe, rather than raw data. They summarize the findings and draw conclusions based on the aggregated information.

Badgley, Catherine, Jeremy Moghtader, Eileen Quintero, Emily Zakem, M. Jahi Chappell, Katia Avilés-Vázquez, Andrea Samulon, and Ivette Perfecto. 2007. "Organic Agriculture and the Global Food Supply." *Renewable Agriculture and Food Systems* 22 (2): 86–108.

Low yields and available organic fertilizer are the two most common arguments against organic agriculture. This article evaluates both claims by reviewing global datasets and modeling the potential food supply that could be grown organically on the current agriculture land base. The authors found that organic agriculture does have the potential to meet the global food needs with the current land base and without requiring additional synthetic fertilizer. The article garnered significant discussion that was published in the editorial of the same issue.

Barański, Marcin, Leonidas Rempelos, Per Ole Iversen, and Carlo Leifert. 2017. "Effects of Organic Food Consumption on Human Health; the Jury Is Still Out!" *Food & Nutrition Research* 61 (1): 1–5.

This article reviews the current literature and analysis that explore the human health impacts of eating organic food. The authors find some areas where there is clear evidence across a number of studies that organic food is nutritionally better than conventionally produced food. They also find many topics that have not been sufficiently studied and that do not have enough data to support the claims that organic is better.

Baron, Haley, and Carolyn Dimitri. 2019. "Relationships along the Organic Supply Chain." *British Food Journal* 121 (3): 771–86.

Very little research has been done on supply chains because of proprietary information restrictions. This study is one of the few that has interview data from certified organic handlers about their relationships to their suppliers. They also report on the results of a wider survey that garnered 153 responses.

Baudry, Julia, Karen E. Assmann, Mathilde Touvier, Benjamin Allès, Louise Seconda, Paule Latino-Martel, Khaled Ezzedine, et al. 2018. "Association of Frequency of Organic Food Consumption with Cancer Risk: Findings From the NutriNet-Santé Prospective Cohort Study." *JAMA Internal Medicine* 178 (12): 1597–1606.

This article reports on one of the largest observational health studies to explore organic food and cancer risk. The findings are similar to one other large study, and they conclude that there is some indication that eating organic food can reduce some cancer risks. This study is not without controversy, and many scientists have indicated that there might be other factors at play.

Bergström, Lars, and Holger Kirchmann. 2016. "Are the Claimed Benefits of Organic Agriculture Justified?" *Nature Plants* 2 (7): 16099.

This is an editorial response to an article by Reganold and Wachter that addressed the sustainability of organic agriculture. The authors of the response claim the sources used to make the argument were biased and did not address nutrients that may be transferred from conventional farms to organic farms through leaching. Reganold and Wachter provided a response to this editorial.

Brennan, Eric B., and Veronica Acosta-Martinez. 2017. "Cover Cropping Frequency Is the Main Driver of Soil Microbial Changes during Six Years of Organic Vegetable Production." *Soil Biology and Biochemistry* 109 (June): 188–204.

Based on data collected from five different management systems on commercial-scale organic vegetable production, this study looked at which system produced more soil microbial action. The number and diversity of soil microbes are important for soil health and are key indicators of soil fertility.

Cambardella, Cynthia A., Kathleen Delate, and Dan B. Jaynes. 2015. "Water Quality in Organic Systems." *Sustainable Agriculture Research* 4 (3): 60.

Water quality in the Midwest is often subpar due to nitrogen leaching into the groundwater. This article summarizes the results of a study completed by the USDA on the potential for organic field crop and pastureland to reduce contamination.

Cernansky, Rachel. 2018. "We Don't Have Enough Organic Farms. Why Not?" *National Geographic*, November 20. https://www.nationalgeographic.com/environment/future-of-food/organic-farming-crops-consumers/.

This article uses striking photographs and personal stories to illustrate the many challenges organic farmers face and why there are not enough organic farmers to meet the demand for domestically grown organic food.

Clark, Michael, and David Tilman. 2017. "Comparative Analysis of Environmental Impacts of Agricultural Production Systems, Agricultural Input Efficiency, and Food Choice." *Environmental Research Letters* 12 (6): 1–11.

This is a meta-analysis of life-cycle assessments of various agriculture systems. The authors argue that based on their assessments switching to plant-based diets would have a more positive impact on the environment than organic agriculture production or grass-fed beef systems.

Crowder, David W., and John P. Reganold. 2015. "Financial Competitiveness of Organic Agriculture on a Global Scale." *Proceedings of the National Academy of Sciences* 112 (24): 7611–16.

This paper scrutinizes datasets for 55 crops worldwide for financial performance of organic versus conventional agriculture systems. They compare systems with and without the addition of price premiums for organic systems and find the breakeven premium needed to make organic agriculture competitive with conventional. They also analyze the cost/benefit ratios of various scenarios.

Delate, Kathleen, Cynthia Cambardella, Craig Chase, and Robert Turnbull. 2015. "A Review of Long-Term Organic Comparison Trials in the U.S." *Sustainable Agriculture Research* 4 (3): 5.

This article provides an overview of six major long-term organic field crop farming systems trials that have been running since the 1980s and 1990s. The authors describe each research location and the major results of the studies conducted there. While they do highlight some overall trends from the six trial sites, they do not attempt to aggregate data for a larger study.

Delate, Kathleen, Stefano Canali, Robert Turnbull, Rachel Tan, and Luca Colombo. 2017. "Participatory Organic Research in the USA and Italy: Across a Continuum of Farmer–Researcher Partnerships." *Renewable Agriculture and Food Systems* 32 (4): 331–48.

On-farm research allows researchers to test methods and applications in a real-world production setting rather than just on a test plot. Many researchers and farmers find working together makes the research more relevant and adaptable and opens up communication between two groups that are not normally talking about production issues. This study compares the collaborative research done in the United States with that done in Italy to identify lessons learned and to share ideas about improving the programs and policies in each location.

DeLind, Laura B. 2000. "Transforming Organic Agriculture into Industrial Organic Products: Reconsidering National Organic Standards." *Human Organization* 59 (2): 198–208.

This article provides a good insight into the concerns organic growers had about a national organic standard and the challenges in making organic farming practices fit a uniform standard. DeLind includes numerous quotes and real-life examples to make the case that organic certification transformed a diverse community of growers into an industry that closely resembles conventional agriculture.

Gomiero, Tiziano, David Pimentel, and Maurizio G. Paoletti. 2011. "Environmental Impact of Different Agricultural Management Practices: Conventional vs. Organic Agriculture." *Critical Reviews in Plant Sciences* 30 (1–2): 95–124.

This paper reviews organic and conventional systems for three major environmental impacts: soil quality, biodiversity, and energy use. It also gives an overview of some of the barriers to adopting organic agriculture methods.

Guthman, Julie. 2004. "The Trouble with 'Organic Lite' in California: A Rejoinder to the 'Conventionalisation' Debate." *Sociologia Ruralis* 44 (3): 301–16.

In 1997, Guthman and several colleagues published an article that sparked a wide-ranging debate on the involvement of agri-business in the organic sector. This article builds on and clarifies their original argument with additional data and discussion. The conventionalization argument coined in their first article is prevalent in the literature now.

Heckman, J. 2006. "A History of Organic Farming: Transitions from Sir Albert Howard's War in the Soil to USDA National Organic Program." *Renewable Agriculture and Food Systems* 21 (3): 143–50.

Following the development of organic agriculture with a special emphasis on Sir Albert Howard's work, this article focuses particularly on how his concepts related to the leading theories on soil chemistry at the time. The author explains how Howard himself had a role to play in polarizing organic and conventional agriculture and how Howard might perceive the modern organic farming culture.

Hemler, Elena C., Jorge E. Chavarro, and Frank B. Hu. 2018. "Organic Foods for Cancer Prevention—Worth the Investment?" *JAMA Internal Medicine* 178 (12): 1606–7.

This is a commentary on the article Baudry et al. published in the same issue. It highlights some of the weaknesses inherent in a study that relies on self-reported food intake. The authors suggest that more research is needed before recommendations for eating organic food are made because the costs of organic may discourage more produce consumption.

Howard, Philip. 2009. "Consolidation in the North American Organic Food Processing Sector, 1997 to 2007." *International Journal of Sociology of Agriculture and Food* 16 (January): 13–30.

This article provides details of the concentration and integration of the organic food sector for a period of 10 years. Highlighting companies and brands acquisitions and mergers, the article uses graphics to show the changing industry structure and the trends that helped bring about these changes.

Howard, Philip H. 2015. "Intellectual Property and Consolidation in the Seed Industry." *Crop Science* 55 (6): 2489–95

Consolidation of the seed industry paired with increasing intellectual property rights being applied to the seed sector produced drastic changes in agriculture, and it has particular implications for the organic sector. This article reviews the policy and structural changes that have occurred and the resulting impacts both in the short term and long term. A discussion of potential solutions is included at the end.

Hyland, Carly, Asa Bradman, Roy Gerona, Sharyle Patton, Igor Zakharevich, Robert B. Gunier, and Kendra Klein. 2019. "Organic Diet Intervention Significantly Reduces Urinary Pesticide Levels in U.S. Children and Adults." *Environmental Research* 171 (April): 568–75.

This study tests urine samples for pesticides before and after an organic diet was imposed for five days. The results confirmed a reduction in pesticide residues after eating organic food. This was a limited study conducted on only 16 participants over a short time span, but the results warrant further study.

Kim, GwanSeon, Jun Seok, and Tyler Mark. 2018. "New Market Opportunities and Consumer Heterogeneity in the U.S. Organic Food Market." *Sustainability* 10 (9): 3166.

This study looks for differences between organic consumers within a particular income level using statistical modeling. They use this information to explore how those differences might impact marketing approaches and product potential and what factors make consumers more likely to continue purchasing organic food.

Kirchmann, Holger. 2019. "Why Organic Farming Is Not the Way Forward." *Outlook on Agriculture* 48 (1): 22–27.

There aren't many scientific articles that directly question organic farming as a whole. Most articles and studies focus on some aspect of organic agriculture that is not working well or does not perform as well as other production methods. The point of this article is to summarize data that supports the author's argument that there is no scientific validity for organic production methods.

Lockeretz, W., G. Shearer, and D. H. Kohl. 1981. "Organic Farming in the Corn Belt." *Science* (New York, N.Y.) 211 (4482): 540–47.

One of the foundational research articles on organic farming published in a peer-reviewed research journal, the authors provided a summary of the major comparison studies on organic and conventional farmers. They looked at factors such as social and economic perspectives, farming practices, crop yields and quality, soil quality, and energy use.

Marasteanu, I. Julia, and Edward C. Jaenicke. 2019. "Economic Impact of Organic Agriculture Hotspots in the United States." *Renewable Agriculture and Food Systems* 34 (6): 501–22.
By first identifying counties with high numbers of organic farm operations, the authors use statistical modeling to determine the economic impact of organic production. They compare these organic hotspots to other agricultural hotspots, and economic indicators for areas not in a hotspot ensure the impact is actually related to the presence of numerous organic farm operations.

Meemken, Eva-Marie, and Matin Qaim. 2018. "Organic Agriculture, Food Security, and the Environment." *Annual Review of Resource Economics* 10 (1): 39–63.
A review article with a very broad scope, this article makes the case that on a global scale organic agriculture cannot meet the needs of the current population while maintaining the current yield levels. It concedes that organic farming systems are beneficial for the environment, but it suggests that these methods need to be paired with more technological solutions.

Mie, Axel, Helle Raun Andersen, Stefan Gunnarsson, Johannes Kahl, Emmanuelle Kesse-Guyot, Ewa Rembiałkowska, Gianluca Quaglio, and Philippe Grandjean. 2017. "Human Health Implications of Organic Food and Organic Agriculture: A Comprehensive Review." *Environmental Health* 16 (1): 111.
This article summarizes the existing data on human health impacts of organic food. They include a review of animal health studies to see if there are any insights for human health, but they did not find any that were

directly related. They detail pesticide exposure, nutrient content, and antibiotic resistance studies that have been done to date.

Muller, Adrian, Christian Schader, Nadia El-Hage Scialabba, Judith Brüggemann, Anne Isensee, Karl-Heinz Erb, Pete Smith, et al. 2017. "Strategies for Feeding the World More Sustainably with Organic Agriculture." *Nature Communications* 8 (1): 1290.

The authors of this study address the issue of low yields and nutrient availability by taking a food systems approach to their analysis. This approach uses factors that many other studies ignore, such as food waste, animal feeding systems, and consumption trends. Rather than trying to compare the production systems currently in place, they look at what factors would have to change in order to make organic agriculture successful on a larger scale. They use advanced modeling techniques to explore a range of options.

Ponisio, Lauren C., Leithen K. M'Gonigle, Kevi C. Mace, Jenny Palomino, Perry de Valpine, and Claire Kremen. 2015. "Diversification Practices Reduce Organic to Conventional Yield Gap." *Proceedings of the Royal Society B: Biological Sciences* 282 (1799): 20141396.

An updated meta-analysis of the yield comparisons between organic and conventional farming systems, this study uses a much larger dataset than any previous studies. In addition, they developed a new framework that better accounts for the diversification and crop rotations found in organic farming systems.

Reganold, John P., and Jonathan M. Wachter. 2016. "Organic Agriculture in the Twenty-First Century." *Nature Plants* 2 (2): 15221.

This article argues that although organic farming systems produce lower yields than conventional farming systems, they perform better on a number of sustainability metrics. This article prompted a letter to the editor (see Bergström and Kirchmann) and a subsequent response from the authors.

Reganold, John P., and Jonathan M. Wachter. 2016. "Reply to 'Are the Claimed Benefits of Organic Agriculture Justified?'" *Nature Plants* 2 (7): 16099.

The authors respond to the critique of their paper by pointing out that they used all available review and meta-analysis papers and specifically did not use single-study papers to support their argument; therefore, they should not be used to refute the article.

Röös, Elin, Axel Mie, Maria Wivstad, Eva Salomon, Birgitta Johansson, Stefan Gunnarsson, Anna Wallenbeck, et al. 2018. "Risks and Opportunities of Increasing Yields in Organic Farming. A Review." *Agronomy for Sustainable Development* 38 (2): 14.

Since lower organic yields are often targeted in arguments against more widespread organic production, these authors reviewed the potential impact of increasing production yields on a range of environmental, social, and economic factors. They used this data to identify which strategies for increasing yields will result in the lowest negative impacts so that resources can focus on those efforts.

Scialabba, Nadia El-Hage, and Maria Müller-Lindenlauf. 2010. "Organic Agriculture and Climate Change." *Renewable Agriculture and Food Systems* 25 (2): 158–69.

This article addresses some of the specific features of organic farming systems that can help mitigate the climate crisis through both lower GHG emissions and by improving soil carbon sequestration.

Seufert, Verena, and Navin Ramankutty. 2017. "Many Shades of Gray—The Context-Dependent Performance of Organic Agriculture." *Science Advances* 3 (3): e1602638.

Using a wider framework of exploration than most, this paper looks at the various contexts in which organic production can be positive or negative using economic, environmental, and social indicators. The paper includes a number of very well-done figures and tables to illustrate the variable results and the importance of context.

Shreck, Aimee, Christy Getz, and Gail Feenstra. 2006. "Social Sustainability, Farm Labor, and Organic Agriculture: Findings from an Exploratory Analysis." *Agriculture and Human Values* 23 (4): 439–49.

One of the few articles that focuses on farmworkers in organic agriculture, this paper looks at social sustainability. The authors surveyed organic farmers in California, where many farms rely on farmworkers for at least some portion of the growing season. They focus on the perspectives of the farmowners and the socially sustainable practices they do or do not implement on their farms.

Simonne, Amarat, Monica Ozores-Hampton, Danielle Treadwell, and Lisa House. 2016. "Organic and Conventional Produce in the U.S.: Examining Safety and Quality, Economic Values, and Consumer Attitudes." *Horticulturae* 2 (2): 5.

This article offers a very succinct overview of the most commonly discussed nutritional and food safety comparisons between organic and conventional agriculture. It summarizes the findings from a number of recent studies and suggests areas where more research is needed.

Skinner, Colin, Andreas Gattinger, Maike Krauss, Hans-Martin Krause, Jochen Mayer, Marcel G. A. van der Heijden, and Paul Mäder. 2019. "The Impact of Long-Term Organic Farming on Soil-Derived Greenhouse Gas Emissions." *Scientific Reports* 9 (1): 1–10.

This is a highly technical article that compares GHG data on multiple farming systems. The discussion section of the article highlights their findings and compares them to similar studies completed by other researchers. Despite being written for a technical audience, it does offer some firm results that indicate organic agriculture's role in the climate crisis.

Smith, Olivia M., Abigail L. Cohen, John P. Reganold, Matthew S. Jones, Robert J. Orpet, Joseph M. Taylor, Jessa H. Thurman, et al. 2020. "Landscape Context Affects the Sustainability of Organic Farming Systems." *Proceedings of the National Academy of Sciences* 117 (6): 2870–78.

Landscape or the way surrounding the land is being used is not often considered in studies on organic agriculture production. This article looks at the impact of various landscapes on a number of sustainability factors, yields, and profitability on organic farms. The authors find fairly strong interplay between landscapes and biodiversity, yield, and profitability.

Tuck, Sean L., Camilla Winqvist, Flávia Mota, Johan Ahnström, Lindsay A. Turnbull, and Janne Bengtsson. 2014. "Land-Use Intensity and the Effects of Organic Farming on Biodiversity: A Hierarchical Meta-Analysis." Edited by Ailsa McKenzie. *Journal of Applied Ecology* 51 (3): 746–55.

Repeated studies have found that organic farms improve species richness by about 30 percent. This article supports that finding, but it delves deeper to explain how surrounding agricultural practices impact biodiversity levels as well as crop type. They identify a number of regions that have been understudied.

Veldstra, Michael D., Corinne E. Alexander, and Maria I. Marshall. 2014. "To Certify or Not to Certify? Separating the Organic Production and Certification Decisions." *Food Policy* 49 (December): 429–36.

Drawing on data from a survey of fruit and vegetable farmers in the United States, the authors explore the decision-making process for organic certification. They attempt to answer questions on why farmers use organic practices, why they choose to certify, and what factors are most likely to impact those decisions. While most articles on farmers' decisions to adopt organic focus on why they choose the farming practices, this article adds to that discussion by separating it from the question of becoming certified.

Vogl, Christian R., Susanne Kummer, Friedrich Leitgeb, Christoph Schunko, and Magdalena Aigner. 2015. "Keeping the Actors in the Organic System Learning: Role of Organic Farmers' Experiments." *Sustainable Agriculture Research* 4 (3): 140.

Organic farmers are always experimenting with different methods and adapting them to their own particular needs and farm ecosystems. Scientists have known that farmers try out new things, but they are rarely captured in academic work. This article draws on interviews with organic farmers in two countries to determine how they go about conducting their on-farm research and look for ways in which it can inform more academic work on innovation in organic agriculture.

Welsh, Jean A., Hayley Braun, Nicole Brown, Caroline Um, Karen Ehret, Janet Figueroa, and Dana Boyd Barr. 2019. "Production-Related Contaminants (Pesticides, Antibiotics and Hormones) in Organic and Conventionally Produced Milk Samples Sold in the USA." *Public Health Nutrition* 22 (16): 2972–80.

This article is a summary of research that uses laboratory analysis of organic and conventional milk to determine the levels of pesticides, antibiotics, and hormones present in each. They purchased milk in nine locations to use in their tests. The article includes detailed tables showing the amounts of contaminants found.

White, Kathryn E., Michel A. Cavigelli, Anne E. Conklin, and Christopher Rasmann. 2019. "Economic Performance of

Long-Term Organic and Conventional Crop Rotations in the Mid-Atlantic." *Agronomy Journal* 111 (3): 1358.

Given the organic grain shortage in the United States, these authors decided to provide an updated analysis on the economics of organic crop production in the mid-Atlantic region. This region has received much less attention for its crop production because it is better known for its dairy farms and diverse vegetable farms. Detailed results are displayed in tables and graphs and discussed in detail.

Youngberg, G., and S. P. DeMuth. 2013. "Organic Agriculture in the United States: A 30-Year Retrospective." *Renewable Agriculture and Food Systems* 28 (4): 294–328.

Authored by one of the members of the USDA study team for organic farming and the only organic staff member the USDA employed until the passage of the OFPA, this report gives an insider perspective to the evolution of organic farming in the United States. It traces both the ideology of organic farming and changes that took place inside the USDA as public perceptions and policy changes shifted the focus on organic.

Reports

Farmworker Justice. 2013. "Exposed and Ignored." https://www.farmworkerjustice.org/sites/default/files/aExposed%20and%20Ignored%20by%20Farmworker%20Justice%20singles%20compressed.pdf.

This report describes the risks that farmworkers and their families face from pesticide exposure on farms. It details the failings of the Worker Protection Standard maintained by the EPA and includes many recommendations for improving farmworker safety.

Greene, Catherine, Gustavo Ferreira, Andrea Carlson, Byce Cooke, and Claudia Hitaj. 2017. "Growing Organic

Demand Provides High-Value Opportunities for Many Types of Producers." Amber Waves. https://www.ers.usda.gov/amber-waves/2017/januaryfebruary/growing-organic-demand-provides-high-value-opportunities-for-many-types-of-producers/.

This report is rich with data presented in numerous graphs, tables, and maps. It provides detailed explanations for the growth of organic products in various parts of the United States and for particular demographics of consumers. It covers product premiums as well as production costs for some products.

Greene, Catherine, Sam Wechsler, Aaron Adalja, and James Hanson. 2016. "Economic Issues in the Coexistence of Organic, Genetically Engineered (GE), and Non-GE Crops." Economic Information Bulletin 149. USDA Economic Research Service. https://www.ers.usda.gov/webdocs/publications/44041/56750_eib-149.pdf?v=0.

GE crops pose a number of challenges for organic farmers. This report brings together data on the use of certified organic non-GE and GE seeds in the United States and the economic implications of each. It reveals hardships and losses that organic farmers experience when GE crops contaminate their seeds or crops and the challenges in detecting and avoiding contamination.

Hubbard, Kristina, and Zystro, Jared. 2016. "State of Organic Seed, 2016." Port Townsend, WA: Organic Seed Alliance.

This is the most recent report in an ongoing project by the Organic Seed Alliance to monitor and share information on the organic seed sector in the United States. Reports will be released approximately every 5 years. The first report released in 2011 shared the results of a sector-wide needs assessment. This second report is the first update and draws on data from four separate surveys and eight additional listening sessions held at organic farming conferences. The report describes the state of organic seed research and seed supply chains and an overview of various policy issues. The report includes a range of recommendations and plans for next steps.

Kuepper, George. 2010. "A Brief Overview of the History and Philosophy of Organic Agriculture." Poteau, OK: Kerr Center for Sustainable Agriculture.

While there are many articles and books about the history of organic farming, most are scholarly or focus on one particular aspect. This report is a concise account of the history of organics up to about 2006. It is straightforward and briefly covers all the important highlights including the creation of the NOP. There is also a neat timeline illustration that shows the multitude of influences at various points and how that led to the current understanding of organics.

Niggli, Urs, A. Fließbach, H. Schmid, and A. Kasterine. 2007. "Organic Farming and Climate Change." Geneva: International Trade Center UNCTAD/WTO. https://orgprints.org/13414/3/niggli-etal-2008-itc-climate-change.pdf.

While the report is outdated, the issues covered are still very relevant. The suggestions the report makes for using organic agriculture to mitigate climate change are still being discussed despite newer data being available. The section on adaptation to climate change and the areas in which organic agriculture needs to improve are still the main issues discussed today. This report makes highly technical information easy to understand.

"Organic Farming Is on the Rise." n.d. Pew Research. https://www.pewresearch.org/fact-tank/2019/01/10/organic-farming-is-on-the-rise-in-the-u-s/.

This interactive online report gives a good overview of the 2016 statistics on organic farming in the United States based on data from the USDA organic survey (hopefully, Pew

Research will update with data from the 2019 survey once it is available). There are some very good graphics to illustrate the spread and change of organic production between 2011 and 2016. They also pull in highlights from other reports they have conducted on organic food consumer perspectives on organic and public perspectives on food risks.

Pittman, Harrison, M. 2004. "A Legal Guide to the National Organic Program." An Agricultural Law Research Article. University of Arkansas: The National Agricultural Law Center.

This report provides a plain language guide to the requirements for becoming certified as organic. It is very useful for understanding the legal issues involved in becoming certified and especially for the steps that can be taken when certification is denied or violations are found.

Reaves, Elizabeth, and Rosenblum, Nathaniel. 2014. "Barriers and Opportunities: The Challenge of Organic Grain Production in the Northeast, Midwest and Northern Great Plains." Sustainable Food Lab.

Grain production is essential for so many other parts of the organic industry, and this report provides context for understanding the major issues that are holding back production growth. For each major barrier, it provides a number of recommended interventions that are clear and practical with suggestions for specific organizations and companies that can help make them happen. The report focuses on practical actions and programs that are already working to address the issues rather than broad policy approaches that may take many years to implement.

Rural Advancement Foundation International-USA. 2010. "National Organic Action Plan (NOAP): From the Margins to the Mainstream: Advancing Organic Agriculture in the US." https://rafiusa.org/docs/noap.pdf.

This report summarizes five years of work with members of the organic community to create a vision for the future of organic agriculture in the United States. It is designed to be a grassroots vision for reaching a number of sustainability goals. The plan includes benchmarks to be met, suggestions of stakeholders who can best make those happen, and plans for tracking progress and documenting success.

Scialabba, Nadia El-Hage, and Caroline Hattam, eds. 2002. Organic Agriculture, Environment and Food Security. Rome: UN FAO.

This publication focuses on environmental issues and the role of organic agriculture. The first half covers topics such as biodiversity, climate change, desertification, and soil fertility as well as the growth of the organic industry on a global scale. The second half focuses on organic agriculture projects and innovations in the developing world. It includes numerous case studies to illustrate the community-level impacts of organic agriculture production and opportunities for creating more food security. The editors draw on a number of scholars who contributed to various parts of the book. It is straightforward and easy to read with lots of figures and tables with relevant data to support their arguments.

Sligh, Michael, and Carolyn Christman. 2003. "Who Owns Organic? The Global Status, Prospects, and Challenges of a Changing Organic Market." Pittsboro: Rural Advancement Foundation International. https://issuu.com/rafi-usa/docs/who_owns_organic.

This report provides a good snapshot of the status of the organic sector shortly after the OFPA regulations came into effect in 2002. It provides an overview of the global numbers of organic acreage and sales along with details about certifications programs across the globe. It is useful for comparisons to the current structure of the organic industry as it provides lots of details about corporate structures and distribution at that time.

Stephenson, Garry, Lauren Gwin, Chris Schreiner, and Sarah Brown. 2017. "Breaking New Ground: Farmer Perspectives on Organic Transition." Center for Small Farms & Community Food Systems Oregon State University and Oregon Tilth. https://tilth.org/app/uploads/2017/03/OT_OSU_Transition-Report_03212017.pdf.

This report shares the results of a national survey of farmers who have been through organic transition, are considering transitioning to organic, or are somewhere in the middle of the process. They go beyond the presentation of aggregate data and break it down into four different categories of farmers and present both data gathered and recommendations for supporting farmers at each stage of transition.

Strauss. 2018. "Americans Are Divided Over Whether Eating Organic Food Makes for Better Health." Fact Tank. Pew Research Center. https://www.pewresearch.org/fact-tank/2018/11/26/americans-are-divided-over-whether-eating-organic-foods-makes-for-better-health/.

This report pulls together data on American consumer perspectives on organic food that PEW collected in a broader survey on public perspectives on food risks. The report discusses the demographics about who are most likely to buy organic food or believe it to be better for health. The report also covers how beliefs about organic food align with other beliefs about food safety.

United States Department of Agriculture Study Team on Organic Farming. 1980. "Report and Recommendations on Organic Farming." Washington, DC: U.S. Department of Agriculture. https://pubs.nal.usda.gov/sites/pubs.nal.usda.gov/files/Report%20and%20Recommendations%20on%20Organic%20Agriculture_0.pdf.

A ground-breaking effort at the USDA, this report is a thorough compilation of the status of organic farming in the 1970s. It characterizes organic farms, farmers, farming practices, marketing strategies, and the known environmental impacts of organic production. It provides scientific analysis of the most common organic farming methods and an overview of the research and educational programs underway at the time. It finishes with a description of the challenges in place at the time and provides recommendations for expanding organic programs at the USDA.

USDA Office of Inspector General. 2005 and 2010. "Agricultural Marketing Service's National Organic Program" & "Oversight of the National Organic Program." Audit Reports 01001-02-Hy & 01601-03-Hy. Washington, DC: USDA Office of Inspector General. https://www.usda.gov/oig/webdocs/01001-02-HY.pdf & https://www.usda.gov/oig/webdocs/01601-03-HY.pdf.

In 2005 and 2010, the NOP underwent audits by the USDA's Office of Inspector General. These reports cover the results of the audits, including several findings that required a response and action taken by the AMS and NOP. The response from AMS and NOP to each of the findings and recommendations is included at the end.

Viña, Stephen R. 2006. "Harvey v. Veneman and the National Organic Program: A Legal Analysis." Congressional Research Service. https://nationalaglawcenter.org/wp-content/uploads/assets/crs/RS22318.pdf.

This is a brief but detailed account of the court rulings in the Harvey versus Veneman court case written for Congress. The report breaks down the three counts that were not dismissed into a summary of court action, congressional action, and administrative action, so readers can see how each branch of government dealt with the court rulings.

Internet Sources

Civil Eats. https://civileats.com.

This is an independent news outlet focused on producing stories that offer critical perspectives on the American food system. Civil Eats maintains a strong focus on sustainable agriculture and economically and socially just communities. They cover issues as wide-ranging as policy, health, environment, and farming. There are many stories concerning organic agriculture, and they often provide a perspective that incorporates the social justice issues that are often lacking in other reporting on the organic sector.

"EOrganic." https://eorganic.info/.
The USDA-run extension services is a long-standing community of professionals that provides services and outreach in rural communities across the United States. eOrganic is a subcommunity of professionals (not just USDA staff) focused on organic farming that has created a databased of peer-reviewed content related to organic farming and food. The website includes access to articles, videos demonstrating organic practices, webinars, and Q & A forums. There are also some short courses available on a few organic farming topics. It contains a lot of practical information on particular problems that organic farmers face and solutions that have worked for farmers. This is also a place to keep up with the activities and resources of many organic farming organizations across the country.

FiBL Organic World: Global organic farming statistics and news. https://www.organic-world.net/yearbook/yearbook-2020.html.
This website is the data collection arm of the Research Institute of Organic Agriculture, located in Switzerland. Each year, they publish a yearbook that documents annual statistics and worldwide trends in collaboration with IFOAM. The yearbook report has been published every year since 2000. The reports are supplemented with an interactive database and infographics for both global and region-specific data.

Gold, Mary V., and Jane Potter Gates. 2007. "Tracing the Evolution of Organic/Sustainable Agriculture: Alternative

Farming Systems." https://www.nal.usda.gov/afsic/tracing-evolution-organic-sustainable-agriculture.

The USDA's National Agriculture Library put together a bibliography that focuses on the organic and sustainable literature held in the library's collection. It contains nearly 200 annotated references that are arranged chronologically from the earliest writings available to 2007. There is also an option to view the bibliography by topic.

Howard, Philip H. "Organic." https://philhoward.net/category/organic/.

Philip Howard is a food systems scholar who has focused on consolidation of food and beverage industries and eco-labels. He has developed unique graphics to illustrate the changes to the food sector in a way that text is unable to convey. On his personal website, he maintains graphics that demonstrate the changes in the organic food industry. His organic processing industry structure graphic shows the changes in ownership and control of organic corporations since 1997. He has graphics that show the distribution and retail structure, acquisitions and mergers, organic farm concentration, and the seed industry.

Minnesota Institute for Sustainable Agriculture. "National Sustainable Agriculture Oral History Archive." https://www.misa.umn.edu/publications/sustainableagoralhistoryarchive.

The oral history archive includes videos, recordings, and transcripts on many leaders in the sustainable agriculture movement that have made a great impact on the organic sector. Of particular interest are the interviews with Roger Blobaum, a consultant who has worked with numerous organizations and coalitions including one that helped lobby for the OFPA; Elizabeth Henderson, an organic farmer and well-known organic activist who has won numerous awards for her work in the organic sector; Mark Lipson, an organic leader who started at the OFRF and went on to work as a Organic and Sustainable Agriculture Policy Advisor at the USDA; Kathleen Merrigan, former Deputy Secretary of Agriculture and the author of the OFPA; Jim Riddle and Joyce Ford, pioneering organic farmers and organic inspectors who have been very active in the organic community for over 30 years;. Bob Scowcroft, the former Executive Director of the OFRF and CCOF; Michael Sligh, the Program Director for the RAFI-USA and organic food policy analyst and lobbyist; and Garth Youngberg, the first organic farming coordinator at the USDA and coauthor of the 1980 USDA study on organic farming.

National Climate Assessment. https://nca2014.globalchange.gov/highlights/report-findings/agriculture.

Produced by a Federal Advisory Committee and supported by over 300 experts, this report summarizes the impacts of climate change on seven sectors of the United States including human health, water, energy, transportation, agriculture, forests, and ecosystems. It also summarizes how the impacts will affect the different regions in the United States. The agriculture section includes an interactive map that shows the changes in season length, dry days, and hot nights, which will drastically change agriculture in the coming years. The report details current effects and projects many of the impacts to come. The report is supported by numerous sources and includes numerous figures and graphs, many of them interactive to show changes over time.

"Organic Agriculture." https://www.organicag.org/.

This website is maintained by the Leopold Center of Sustainable Agriculture at Iowa State University. It started as a literature review of scientific articles that compared organic and conventional food, but it has since expanded to other related topics. There are short summaries of each article and a link to the original. It does contain a wealth of peer-reviewed articles, but it does not appear to have been updated since 2015.

"Organic Agriculture: Official Journal of The International Society of Organic Agriculture Research." n.d. https://link.springer.com/journal/13165/volumes-and-issues.

Organic research has come a long way, and it is noteworthy that there is now an academic journal dedicated to organic agriculture. This is an international multidisciplinary journal with an editorial board from universities around the world. The journal has published four issues a year since 2011 on all aspects of production and food systems issues related to organic agriculture. Most of the articles require institutional access or subscription, but a number of them have open access.

Organic ePrints. https://orgprints.org/.

Operated by the International Centre for Research in Organic Food Systems, Organic eprints is an archive of open access electronic documents related to organic food and farming research. It began in 2002, and it has editors who oversee the content from 26 different countries. Resources are in a variety of languages, although most are in English, and are a mix of academic articles, theses, presentations, and reports. The vast majority of the resources are from European sources and not specifically focused on the United States. It is most useful for finding information about organic research and policy from an international perspective.

Pesticide National Synthesis Project. https://water.usgs.gov/nawqa/pnsp/usage/maps/.

The U.S. Geological Survey runs a National Water-Quality Assessment Project that includes the Pesticide National Synthesis Project. They publish tables, graphs, and maps that show estimates of agriculture pesticide use reported by pesticide type. There are also a number of articles and reports on the connection between agricultural pesticides and water quality. The data is available at the state and county level from 1992 to 2017. The map portion is a useful tool for visualizing the increased use of particular pesticides such as Glyphosate by location and crop.

"Sustainable Food Trade Organization." www.sustainablefood trade.org.

This organization offers a number of resources on how organic food companies are working to incorporate a wide range of sustainability issues into their business practices. They publish sustainability progress reports that aggregate data from all of their members plus individual member business sustainability reports.

"USDA Organic Integrity Database." https://organic.ams .usda.gov/Integrity/Reports/Reports.aspx.

The USDA Organic Integrity Database is a searchable database of certified organic operations as compiled by USDA-Accredited Certifying Agents. The reports section of the website contains summary reports of the number of certified organic operations by country and state. There is also a map that shows the concentration of operations across the United States. Historical data going back to 2010 is available on the data history page. Reports can be generated by product, acres, and a number of other signifiers.

Films

Chester, John. 2018. *The Biggest Little Farm*. FarmLore Films: USA. https://www.biggestlittlefarmmovie.com/.

Extraordinarily beautiful cinematography makes this documentary worth watching. The film is about a couple who move to a California property and begin what they call a traditional farm. While the film never mentions the word organic, they are in fact certified organic, and the film demonstrates many practices that are at the foundation of organic production methods. The film shows the evolution of the farm over eight years, which shows just how long it takes farmers to create a balanced ecosystem and all of the learning that must take place to achieve it.

Kiplin, Pastor. 2012. *In Organic We Trust.* Pictures, Incuse. www.InOrganicWeTrust.org.

This documentary film gives a good basic description of organic farming philosophy versus certified organic. It follows one man's quest to understand what organic food is and what it is not. It includes interviews with farmers, organic certifiers, researchers, activists, and those working in all aspects of the food system. The first half focuses on organic production and how it is different from conventional farming and how the use of organic as a marketing tool has changed the movement in the United States. The second half focuses on healthy food and farming solutions in general, including farmers' markets, community gardens, and school food programs.

Smith, Miranda. 1995. *My Father's Garden.* Bullfrog Films: Reading, PA. http://docuseek2.com/bf-mfg.

A moving documentary about two very different farms: an orange grove in Florida and a large crop farm in North Dakota. The filmmaker's father, Herbert Smith, created an orange grove in California during the 1950s and embraced the new chemical farming innovations of the time. Fred Kirschenman returned to his family farm at a time when farmers were in crisis. The film tells the story of his shift to organic farming by following the seasons and demonstrating the differences between his farm and that of his conventional neighbors.

7 Chronology

Organic agriculture developed within a historical context of social change, technological advancement, and transformation in food and farming industries. This chapter provides a map of the major developments, publications, laws, and events that have impacted the evolution of organic food and farming in the United States.

1840 Justus von Liebig builds on the work of Carl Sprengel and proposes the use of inorganic nitrogen-based fertilizers as a replacement for composted manure. He publishes the book *Organic Chemistry in Its Application to Agriculture and Physiology*, which influences others researching agricultural practices to take a reductionist approach.

1862 The United States Department of Agriculture (USDA) is established by President Abraham Lincoln.

1886 Hydrogen cyanide is used as a fumigant on California citrus to prevent damage from scale insects.

1892 Lead arsenate is used to control gypsy moths in apple orchards in Massachusetts.

1910 Federal Insecticide Act is passed by Congress to protect farmers from deceptive labeling on pesticides and insecticides.

Organic strawberries growing in a straw mulch. Studies have shown that conventional strawberries consistently have high levels of pesticide residues. (iStockPhoto)

1911 American agronomist F.H. King publishes his book *Farmers of Forty Centuries: Permanent Agriculture in China, Korea, and Japan*, which details what he learned about soil fertility and farming practice while touring in Asia. His work later inspires J.I. Rodale and other organic pioneers.

1921 Airplanes are used to apply pesticides on crops for the first time.

1924 Rudolf Steiner, founder of the Anthroposophical Society, offers a series of lectures on working in harmony with natural cycles and maintaining a view of the farm as a single system. His ideas become the basis of the biodynamic farming movement.

1924 Albert Howard, a botanist, becomes the director of the Institute of Plant Industry at Indore, India. Albert and his wife Gabrielle, also a botanist, research composting techniques that become known as the Indore Method.

1938 The USDA's annual yearbook is titled *Soils and Men: Yearbook of Agriculture 1938*. It serves as a manual for organic farming that is still referenced today.

1938 DDT is discovered to have insecticidal properties by Swiss scientist Paul Müller.

1939 Lady Eve Balfour begins the Haughley Experiment to compare organic farming techniques with chemical farming on two adjacent farms in England. The experimental farms are eventually taken over by the Soil Association and continue for 30 years.

1940 Lord Northbourne publishes a book called *Look to the Land* in which he coins the term "organic." The term is later adopted by most organic advocates.

1940 *An Agricultural Testament* by Sir Howard Albert is published based on his research in India and goes on to be considered a foundational work in organic agriculture and soil fertility.

1942 J.I. Rodale begins publishing the *Organic Farming and Gardening Magazine* and names Albert Howard as an editor.

1943 Lady Eve Balfour publishes her book *The Living Soil* based on her comparison study of organic and conventional agriculture and inspired by the work of Lord Northbourne and Albert Howard. She goes on to cofound the Soil Association, the main organic advocacy organization in the United Kingdom.

1945 DDT is commercially available for use as an agricultural and household pesticide.

1946 Paul and Betty Keene start Walnut Acres, an organic farm and mail-order food business. Walnut Acres is one of the earliest branded organic food manufacturers.

1946 Farmers begin to use organophosphate insecticides on U.S. farms.

1947 Federal Insecticide, Fungicide, and Rodenticide Act (FIFRA) is passed by Congress. The act requires manufacturers to register their products before selling across state lines and establishes further labeling requirements.

1948 American novelist and writer, Louis Bromfield, publishes *Malabar Farm*; the book contributes to a romanticizing of sustainable and organic farming.

1959 J.I. Rodale publishes the first edition of *The Encyclopedia of Organic Gardening*, a practical handbook still in use today.

1961 *Bacillus thuringiensis* (Bt) is first registered as a pesticide. The pesticide targets the European corn borer, a caterpillar that damages corn crops, but it is also toxic to many other insects including moths and butterflies. The pesticide is eventually genetically engineered into the corn plant.

1962 Rachel Carson publishes *Silent Spring*, a book documenting the devastating impacts of agricultural chemicals. The book sparks an environmental movement.

1970 Environmental Protection Agency (EPA) is created under the Nixon administration in response to the growing concerns over pollution. The agency is now responsible for all pesticide regulations.

1971 Rodale Press is the first organization to begin offering certification of organic farms.

1971 Maine Organic Farming and Gardening Association (MOFGA) is initiated at a meeting organized by Charlie Gould, a University of Maine Cooperative Extension agent. Following the Rodale Organic Garden certification standards, they begin their own certification program in 1972.

1971 The Northeast Organic Farming Association of Vermont forms and creates an organic certification program.

1971 U.S. Secretary of Agriculture Earl Butz infamously states, "Before we go back to organic agriculture somebody is going to have to decide what 50 million people we are going to let starve."

1972 FIFRA becomes the Federal Environmental Pesticide Control Act administered by the EPA. The act strengthens the registration process and allows the EPA to cancel or suspend pesticides that cause unreasonable adverse effects on the environment.

1972 DDT uses for all crops are banned by the EPA.

1972 The International Federation of Organic Agriculture Movements (IFOAM) is created in Versailles, France, during an international meeting on organic agriculture organized by the French organization, Nature et Progrès.

1972 Gene Kahn founds Cascadian Farm in Washington, becoming one of the first to sell processed organic food, starting with frozen fruits and jam. He later sells out to General Mills.

1973 A handful of organic farmers in California create the California Certified Organic Farmers (CCOF) agency and certify 54 farmers in the first year.

1973 The Forschungsinstitut für biologischen Landbau (FiBL), an organic agricultural research institute, is founded in Switzerland.

1974 Oregon Tilth is formed to support organic farmers and gardeners, and it creates a certification program in 1982.

1974 Monsanto brings the glyphosate herbicide to market under the trade name Roundup. The herbicide is used to control broadleaf weeds and grasses.

1975 Masanobu Fukuoka, a microbiologist in Japan, publishes *One Straw Revolution* detailing his experiences in natural farming and a warning against chemical farming. The book is translated into English and inspires farmers around the world to try organic farming techniques.

1975 At least, 11 organizations in the United States are offering organic certification.

1976 New York State becomes the first state to pass legislation allowing for the regulation of organic food.

1977 Wendell Berry publishes *The Unsettling of America: Culture and Agriculture*, a critique of industrial agriculture. He also writes for Rodale Press, including for the magazine *Organic Gardening and Farming.*

1978 The USDA publishes *A Bibliography for Small and Organic Farmers: 1920–1978* compiled by James Schwartz.

1978 EPA releases the first list of registered pesticides with restricted use.

1979 California passes the California Organic Food Act. The law defines organic farming practices but includes no provisions for enforcement or support.

1980 IFOAM develops a set of baseline standards for certification of organic food. The standard is created to help harmonization of standards for global trade.

1980 Whole Foods Market opens in Austin, Texas, one of only a handful of natural food supermarkets. It grows to nearly 500 locations in the United States and United Kingdom.

1980 The USDA publishes the *Report and Recommendations on Organic Farming*, a comprehensive survey of organic agriculture in the United States and recommendations on using organic techniques to improve agriculture. The Reagan administration ignores the report and eliminates the position of Organic Resources Coordinator, marking the beginning of a decade-long hostility toward organic agriculture from the USDA.

1981 Dale Coke, a California vegetable farmer, starts selling spring salad mix to high-end restaurants. Spring salad mix is now a top-selling organic product. He also comes up with the idea of washing the greens in an old washing machine set up to run on the rinse cycle. Many small farmers still use this method of washing greens.

1981 William Lockertz, Daniel Kohl, and Georgia Shearer publish an article in *Science* magazine titled "Organic Farming in the Corn Belt."

1981 American Society of Agronomy hosts a symposium on organic farming and publishes "Organic Farming: Current Technology and its Role in a Sustainable Agriculture."

1982 Representative Jim Weaver of Oregon introduces the Organic Farming Act of 1982, requiring the USDA to create a pilot research program on organic farming techniques. The bill does not gain the support it needs and fails to pass into legislation.

1983 Stonyfield organic dairy farm is founded by Gary Hirshberg and Samuel Kaymen. Ten years later, the company becomes the first to pay dairy farmers not to use synthetic growth hormones.

1983 Austria becomes the first country to establish national regulations for organic farming.

1984 Earthbound Farm is started on 2.5 acres in California by Drew and Myra Goodman. Over two decades, the farm grows to be the largest producer of organic salads in the United States. The farm is acquired by Whitewave Foods in 2013.

1985 The Organic Crop Improvement Association (OCIA) is created in Albany, NY, a farmer networking group modeled after Depression era crop improvement associations. The organization eventually becomes a leading global certification agency.

1985 The Organic Foods Production Association of North America is formed in an effort to have one certification body to meet the needs of distributors, retailers, and processors who are crossing state or regional boundaries. The name changes to the Organic Trade Association (OTA) in 1994.

1988 Congress funds a new competitive grants program called Low-Input Sustainable Agriculture. The name later changes to Sustainable Agriculture Research and Education.

1988 Washington becomes the first state to establish organic standards and certification.

1988 Coulee Region Organic Produce Pool (later named Organic Valley) is formed in Wisconsin. It becomes the largest farmer-owned organic cooperative.

1989 The television program *60 Minutes*, aired by CBS, produces a program called "Intolerable Risk: Pesticides in our Children's Food." The show exposes a report by the Natural Resources Defense Council on the risks of the chemical Alar (Daminozide), a growth regulator sprayed on apples. The public outcry prompts significant consumer interest in organic food.

1989 Elliot Coleman publishes *The New Organic Grower*, a small-scale organic growing manual that becomes a classic text.

1990 Organic sales reach $1 billion in the United States. Sales increase at a rate of 20 percent for the next decade.

1990 Congress passes the Organic Foods Production Act (OFPA) as part of the Farm Bill. The bill authorizes creation of the National Organic Program and the National Organic Standards Board under the USDA Agriculture Management Service.

1991 The European Union adopts regulations for organic production and labeling of plant products.

1991 International Organic Inspectors Association is formed and led by Jim Riddle.

1992 The inaugural National Organic Standards Board (NOSB) is formed. NOSB is a federal advisory board made up of 15 representatives from the organic community. They meet twice a year and provide recommendations to the USDA.

1992 The Organic Farming Research Foundation (OFRF) is formed to support the education and on-farm research of organic farming. Bob Scowcroft is the first Executive Director. The foundation completes the first National Organic Farmers Survey.

1993 The OFRF organizes the first Organic Leadership Conference in Berkeley, California.

1996 Monsanto brings Roundup Ready soybeans to market. By 2005, they have developed Roundup Ready cotton, corn, canola, alfalfa, and sugar beets. The crops are genetically engineered to allow them to tolerate the herbicide glyphosate, the active ingredient in Roundup.

1997 The USDA publishes the first set of organic standards based on recommendations from the NOSB and receives unprecedented feedback from over 275,000 individuals, who were angered to see the "big three"—irradiation, GMOs, and sewage sludge included in the rules. The USDA withdrew the rules and revised them for a final rule published five years later.

1997 Organic Materials Review Institute (OMRI) is formed as a national nonprofit organization to provide independent evaluation of products intended for use in certified organic production. OMRI produces a list of products approved for use under certification and provides technical reports to the USDA.

1997 Mark Lipson, a senior policy analyst with the OFRF, publishes *Searching for the "O-word": Analyzing the USDA Current Research Information System for Pertinence to Organic Farming*. The report documents the status or lack thereof of organic farming research supported by the USDA.

1997 Iowa State University creates the first organic specialist position at a land grant university.

1998 Ronnie Cummins and Rose Welch found the Organic Consumer's Association to harness the momentum of consumer interest in protecting organic integrity after the strong response to the USDA's initial rules.

1999 A joint commission of the FAO and WHO called Codex Alimentarius approves guidelines for the production, processing, labeling, and marketing of organic food.

1999 Minnesota passes legislation providing for an organic certification cost-share program.

1999 Restaurant Nora, located in Washington, DC, works with the certifier Oregon Tilth and becomes the first certified organic restaurant.

2000 A decade after the OFPA is passed, the National Organic Standards are finally published in the Federal Register, establishing the USDA standards for organic food production.

2000 A USDA ERS report *US Organic Farming in the 1990s: Adoption of Certified Systems* by Catherine Greene indicates that organic cropland more than doubled in the past decade.

2001 OFRF publishes *State of the States: Organic Systems Research and Land Grant Institutions* written by Jane Sooby.

2002 The USDA Certified Organic standards and labeling goes into effect. The term "organic" can only be used if a product is certified by an accredited third-party.

2002 Farm Bill includes funding to create an Organic Agriculture Research and Extension program. The program is allocated $15 million to support research over three years.

2003 The National Organic Coalition forms to lobby in Washington on behalf of the organic sector. It is formed as an alliance of organizations from all sectors of the organic movement.

2003 The Organic Seed Alliance is formed to support organic seed availability, research, and policy.

2004 The USDA's Cooperative State Research, Education, and Extension Service allocates $4.7 million to a new Integrated Organic Program. The program funds both the Organic Transitions Program and the Organic Agriculture Research and Extension Initiative.

2005 A blueberry farmer, Arthur Harvey of Maine, wins an appeal against the USDA. He argued that the USDA's final rules were inconsistent with the OFPA. The court agreed on three points: nonorganic ingredients not commercially available in organic form must have individual reviews to be allowed, synthetic substances are not allowed in processed organic foods, and dairy herds in transition must be fed 100 percent organic feed.

2006 Beekeepers note significant colony losses and express concern about the effects of neonicotinoids, a class of insecticides.

2006 The USDA National Library publishes "Organic Farming and Marketing: Publications from the United States Department of Agriculture, 1977–2006" written by Mary Gold and Rebecca Thompson. The following year, Mary Gold ranks the top 10 research journals related to organic food and farming from 2004 to 2007.

2008 California farmer Larry Jacobs wins a pesticide drift case against Western Farm Services, Inc. and is awarded $1 million in damages. The precedent setting case found that contamination of organic crops by pesticides evaporating after application violated the rights of the organic growers even if no laws were broken.

2008 The USDA National Agricultural Statistics Service conducts their first organic survey as part of the agricultural census.

2008 Farm Bill allocates $78 million to continue organic agriculture research and extension.

2009 Kathleen Merrigan is named Deputy Secretary of Agriculture. Kathleen Merrigan was a congressional aide who wrote the federal Organic Food Production Act.

2009 First Lady Michelle Obama installs an organic vegetable garden at the White House, and Secretary of Agriculture Tom Vilsack installs the first organic garden at the USDA headquarters.

2009 Hundreds of organic farmers hold a rally in Wisconsin to put a spotlight on the glut of illegal organic milk coming from large-scale operations that threaten dairy farmers' livelihoods. Secretary of Agriculture Tom Vilsack attends and promises to fully enforce the rules of the National Organic Program.

2010 Following many years of discussion, input, and advocacy work, the NOP publishes the "Pasture Rule." The rule governs the management of ruminate livestock's access to pasture, feed, and grazing.

2010 There are thirty-eight states with rules and regulations relating to organic food and farming.

2010 Organic sales reach $26.7 billion in the United States.

2012 According to a study by the Hartman Group, a consumer research firm, 75 percent of U.S. consumers had used organic products in the past three months. This was in part because organic foods are increasingly found at mainstream grocery stores and are more affordable.

2013 Organic acreage in the United States is over 5 million acres, up from approximately 1 million in 1992.

2015 The USDA proposes an amended version of the Origin of Livestock rule that will close loopholes that allow conventional dairy cows to be added to a herd. They receive 1500 comments supporting the change.

2016 The Organic Farmers Association forms as a farmer-led national advocacy platform. It is supported by the Rodale Institute.

2016 The USDA's Risk Management Agency develops a Whole-Farm Revenue Protection insurance product, allowing diversified farms to gain access to insurance coverage historically only available to commodity producers.

2016 Organic farmers protest outside a NOSB meeting in Stowe, VT, over the organic certification of hydroponic foods.

2016 Total sales of organic products reach over $47 billion in the United States, which is about 4 percent of all food sales.

2017 Amazon.com acquires Whole Foods Market for $13.7 billion. The merger eliminates regional purchasing, which limits small-scale brands' ability to access the market.

2017 NOSB votes to allow hydroponic and aquaponic foods to be certified organic but not aeroponics. The USDA, however, decides to continue allowing all three production methods to be certified organic.

2017 The USDA publishes an Organic Livestock and Poultry Practices Rule requiring outdoor access for all organic chickens. The decision is repealed by the Trump Administration the following year.

2017 An article published in *Science*, "Country-specific effects of neonicotinoid pesticides on honeybees and wild bees," confirms that neonicotinoids are harming bees.

2018 Walmart introduces a line of Wild Oats organic pantry items with no price premium.

2018 Several farmers are charged in the largest case of organic grain fraud in the history of organic certification. The leader, Randy Constant, sold over $140 million dollars of conventional grain as organic over 14 years.

2018 Farm Bill includes historic levels of funding for organic research and education, data collection, cost-share program, additional training for the USDA staff, and improved enforcement authority by the NOP.

2018 Organic dairy farmers experience drastically declining prices for milk, and many are unable to stay in business.

2019 The USDA reopens the Origin of Livestock Rule for a 60-day comment period and receives 600 comments supporting the rule.

2020 The Center for Food Safety along with organic farmers and supporting organizations files a lawsuit against the USDA, claiming hydroponic production does not meet the soil building requirements in the organic standards.

2020 A U.S. District Court finds the USDA used a flawed analysis as the basis for withdrawing the Organic Livestock and Poultry Practices Rule and orders the USDA to fix the errors within 180 days.

2020 The USDA publishes a proposed Strengthening Organic Enforcement Rule designed to implement provisions from the 2018 Farm Bill and recommendations from the NOSB.

2020 The COVID-19 pandemic drives a significant increase in consumer demand for local organic food.

This chapter defines a number of technical terms that are commonly used to describe agricultural production, organic farming, and organic policy issues.

Agroecology The application of ecological principles to agricultural practices by working with natural cycles rather than against them.

Beneficial Organism Any organism that benefits the growing process of a plant by aiding with pest control, pollination, and soil fertility.

Biodiversity The variety of life found on earth or variation of genetics, species, and ecosystems in a particular area.

Biodynamic An approach to farming rooted in the work of Dr. Rudolf Steiner and focused on viewing the farm as an integrated living organism with many parts.

Biological Control The process of regulating or controlling pest populations by using beneficial organisms either by purposely putting them among the crop or by creating habitats that support native populations of beneficial organisms.

Biopiracy Taking control of biological material such as plant genetics and claiming ownership through patents in order to restrict and profit from their use.

CAFO Confined or concentrated animal feeding operations are farms with large numbers of livestock or poultry that are housed indoors or in a small space for at least 45 days or more a year.

Codify The process of turning a set of legal requirements into statutory laws.

Commodity Crops Agricultural crops grown in large volumes that are also nonperishable, easily stored and transported, and are generally sold in a commodities market. The most common commodity crops in the United States include corn, soybeans, and wheat, which are often used to feed livestock or processed into food.

Compost A mixture of decomposed organic matter such as plant material or manure that is used for fertilizing soil.

Cover Crop A plant that is used to protect and enrich the soil, help control weeds or pests, and improve biodiversity rather than be harvested. The plants are often plowed into the soil before reaching maturity.

Crop Diversification Growing a number of different plants or varieties to add economic and ecological resilience to a farm.

Crop Rotation Growing a series of different crops in a succession pattern on a plot of land in order to improve soil fertility.

Drift The physical movement of a prohibited substance from the intended location to another site, usually by force of wind.

Fallow Rotation Land that is left unplowed and unseeded for a growing season or more.

Federal Register The official publication used by the federal government to publish rules and regulations, proposed rules, executive orders, and other notifications from federal agencies.

Forage or Fodder A food made from the whole plant such as hay or corn and used to feed livestock. It can be left standing in the field or cut, dried, and stored.

Genetic Engineering Modifying genetic material to achieve specific desirable traits in a living organism.

Green Manures A crop grown specifically to fertilize the soil and plowed into the soil while it is still green. They often double as cover crops.

Greenhouse Gases Gases that trap heat from the sun causing earth's temperatures to rise. The most common greenhouse gases are carbon dioxide and methane.

Growth Hormones A hormone produced by the pituitary gland that stimulates the growth of cells in animals. The recombinant bovine growth hormone is made in the lab using genetic engineering and administered to cows to manipulate reproduction and lactation.

Humus The organic material in soil that contains nutrients, holds water, and stabilizes soil structure.

Hybrid Seeds A plant created by crossing two different varieties of the same plant. This is accomplished by taking pollen of one variety and manually transferring it to a different variety. The seeds of a hybrid plant will not reproduce true to type.

Hydroponic An agricultural practice that involves growing plants without soil and using liquid nutrients instead.

Inputs Agricultural inputs are products that are purchased off the farm to use in production. These can include equipment, feed, seeds, energy, chemicals, nutrients, compost, etc.

Integrated Livestock Production A farming system that incorporates both animal and crop production so that they are mutually supported and reduce the need for inputs.

Intellectual Property Rights A category of property that includes intangible goods or creations that can be protected by patent, trademarks, or other legal rights.

Intercropping A system of growing that includes growing a plant of one kind in between the rows of another.

Monoculture The practice of growing a single crop in a field at a time, often repeated in a short cycle with one other crop.

Nutrient Cycles The process of moving nutrients from the physical earth into a living organism and back again.

Open-pollinated Varieties When these seeds are isolated from other varieties and allowed to reproduce, they will produce plants that are very similar to the parent plant.

Organic Matter The residue of any recently living organism that is decaying. Also, it's considered to be a mixture of living organisms that break down organic residues, material undergoing breakdown and decay, and already decomposed organic material.

Pulse A crop that fixes nitrogen in the soil and can be harvested dry, such as beans, peas, and legumes.

Soil Biota All of the living organisms that exist in soil.

Soil Fertility The capacity of a soil to sustain life and provide the necessary nutrients needed for growth.

Supply chain The steps and networks that are required to get a product in its original state from the farm to the consumer in its final form.

Synthetic Something that is made from a chemical synthesis rather than a natural process.

Tillage Mechanical preparation of the soil for planting that can be done by hand with shovels, rakes, and hoes or by machine with plows, discs, and harrows.

Vertical Integration Taking multiple stages of a supply chain or production and consolidating them in one company.

Weeds Any plant that is growing where it is not wanted.

Yields A measure of agricultural product grown or raised per unit area of land.

About the Author

Shauna M. McIntyre, PhD, is a freelance writer and an editor. She is a contributing author to the *Encyclopedia of Organic, Sustainable, and Local Food* and *America Goes Green: An Encyclopedia of Eco-Friendly Culture in the United States*. She was formerly the executive director of the Ecological Farmers of Ontario and served on the board of directors for the Organic Council of Ontario. McIntyre received her doctorate from the University of Guelph, where her research focused on local food systems. She earned a master's degree from Southern Illinois University-Carbondale. Her master's research on state support for organic agriculture was published in the *Journal of Sustainable Agriculture*.